住宅新景观
NEW RESIDENTIAL
Landscape

深圳市艺力文化发展有限公司 编

图书在版编目（CIP）数据

住宅新景观 / 深圳市艺力文化发展有限公司编. -- 武汉：华中科技大学出版社，2013.8
ISBN 978-7-5609-9367-6

Ⅰ. ①住… Ⅱ. ①深… Ⅲ. ①住宅－景观设计－世界－图集 Ⅳ. ① TU241-64

中国版本图书馆CIP数据核字（2013）第213834号

住宅新景观

深圳市艺力文化发展有限公司 编

出版发行：华中科技大学出版社（中国·武汉）
地　　址：武汉市武昌珞喻路1037号（邮编：430074）
出 版 人：阮海洪

责任编辑：张雪姣	责任监印：张贵君
责任校对：简晓思	装帧设计：黄秋莹

印　　刷：深圳市汇亿丰印刷包装有限公司
开　　本：965mm×1270mm　1/16
印　　张：24
字　　数：346千字
版　　次：2013年11月第1版　第1次印刷
定　　价：378.00元（USD 75.99）

投稿热线：(027)87545012 design_book_wh01@hustp.com
本书若有印装质量问题，请向出版社营销中心调换
全国免费服务热线：400-6679-118 竭诚为您服务
版权所有　侵权必究

PREFACE
序 言

Along with the rapid economic development, the faster urbanization moves on, the higher requirements citizens raise for the quality of the vital living environment. Hence, the issue of landscape design has received more and more concern during the rapid development of urbanization. Residential landscape design, as the main factor affecting the dwelling environment, has an unparalleled effect in improving citizens' life quality and improving the quality of environment.

The landscape design in residential districts often plays a decisive role in the districts' quality. The residential landscape's style often needs to be closely connected to the styles of its surrounding architecture. For a good residential landscape design, designers have to understand the styles of the architecture. From a professional point of view, the residential landscape is a kind of environmental coordination with the architecture in the district. Thus, the landscape design is attached to the architectural design. Its style cannot digress from the architectural styles. Otherwise, the landscape and the architecture will be separated from each other instead of being integrated naturally.

The ultimate goal of landscape design is to serve the residents. It should pay more attention to the sustainability, economical efficiency, practicability and rationality of the environment while planning a landscape design. At the same time, it should lay emphasis on the combination of its greenery, water, surrounding and so on, with function, landscape and culture as a whole, to create a three-dimensional natural space where there is a kind of horizontal and vertical integration. It is of rich ornamental value and economic benefits while meeting the residents' high requirements to the environment, making the resident feeling both mentally and physically comfortable.

This book brings together a great number of representative residential landscape designs, collects the residential scenery all around the world. It offers the readers the residential landscapes worldwide, both in the communities and in the private houses, meanwhile a visual feast of residential landscapes.

伴随着社会经济的高速发展，城市化进程越来越快，人们对赖以生存的环境的质量提出了越来越高的要求。因此，在城市化迅猛发展的过程中，景观设计问题愈来愈受到人们的重视。而住宅景观设计作为影响人们居住环境的一个主要因素，在提高人们生活质量、改善环境质量等方面都是其他因素无法比拟的。

居住区的品质如何，其中的景观设计往往起着决定性的作用。而居住区的景观风格定位，往往需要与其周边的建筑风格密切联系起来，能否把住宅景观设计做好，设计师应对住宅区的建筑风格有一定把握的解读。从专业的角度而言，住宅景观是对住宅区域建筑的环境配合，因此，景观设计是附属于建筑设计风格的，景观风格不能独立于建筑风格之外另起炉灶，否则就会使建筑与景观互相分离，不能自然地融合在一起。

景观设计的最终目的是为人服务，在规划住宅景观时应更多地考虑环境的可持续性、经济性、实用性以及合理性，把设计风格在真正融入实际景观中，同时注重绿脉、水脉、文脉等的结合，并融功能、景观、文化于一体，以创造一种横向、纵向一体化的立体自然空间，使其在富有观赏价值及符合经济效益的同时，还能满足居住者对环境提出的高要求，使居民感觉身心舒畅。

本书集合了大量具有代表性的住宅景观设计案例，收藏世界各地住宅风景，从社区住宅及私人住宅两大方向带领读者品鉴世界各地的景观特色，享受住宅景观的视觉盛宴。

CONTENTS 目 录

Residential Community 居民小区

008	Larkspur Courts 拉克斯普庭院	102	Oak Bay Stage II-III 橡树湾 II-III 期
014	Kingkey Oriental Grace 京基御景东方	106	Shanghai Greentown Rose Garden 上海绿城玫瑰园
020	Vanke Rancho Santa Fe II 万科兰乔圣菲二期	116	Long Beach Villas 长堤花园别墅
026	Kingold HuiJing New Town - LongXi Mountain Park 侨鑫汇景新城——龙熹山山顶公园	124	Toscana 托斯卡纳社区
032	Times Times Peanut 时代·时代花生	132	Vanke Rancho Santa Fe 万科兰乔圣菲
036	Zhuhai Nanfu Jinyuan Landscape 珠海南福锦园景观设计	138	Conghua Hot Spring Villas 从化温泉别墅
042	Stanford West Apartments 斯坦福西区公寓	146	Magee Ranch 麦琪大农场
050	Hangzhou - Spring River Flower Moon 杭州——春江花月	152	Verakin New Town 同景新城
058	Dakota Residences 达科塔住宅	160	Vanke Longgang Mountain Living 万科龙岗山城小区
064	Zobon City Sculpture Garden 中邦城市雕塑花园	168	Nanshan Suzhou Golden Garden 1958 南山苏州金色花园 1958
070	Mantra Aqua Resort Mantra Aqua 度假村	172	Chevron Garden 雪佛龙花园
076	Circle on Cavill 卡维尔之圈	174	Villas del Mar at Palmilla 德尔玛别墅
080	The Trillium 延龄草社区	180	Sierra Bonita Mixed Use Development Sierra Bonita 综合大楼
084	Bamboo Garden 竹园	184	South Park Streetscape and Mixed Use Development 南园街景和综合高层大楼
088	Shanghai Greentown 上海绿城	188	Lander Taizhou LaiYinDongJun 莱茵达泰州莱茵东郡
092	Yanlord Riverside Home St III 仁恒河滨花园第三期	190	Kuany – Huizhou Holland City 光耀——惠州荷兰小城
096	Riverlight Apartments Riverlight 公寓		

192	Campbell, Salice & Conley Residence Halls 坎贝尔、萨利斯 & 康利学生宿舍楼	220	Xinsheng Oriental Blessing 杭州欣盛东方福邸景观设计
196	California State Polytechnic University, Pomona 加州州立理工大学住宅区，波莫纳	228	Southport Central 南港中心
200	UCLA Northwest Campus Redevelopment UCLA（加州大学洛杉矶分校）西北校区重建项目	230	Cutters Landing 卡特斯码头
208	UC Davis West Village Phase I 戴维斯大学西村第一期	234	Halcyon Waters 宁静水域
216	Lantern Bay Master Plan and Coastal Design 灯笼湾总规划和景观设计	238	Kunming Horti-Expo Eco-Communities 昆明园博生态社区

Private Residence 私人住宅

246	French Residence French 住宅	318	Between Predictable & Unexpected 预见与意外之间
252	LeKander Residence 利坎德之家	322	Moltz Landscape 墨尔茨景观
260	James Street Garden 詹姆斯大街花园	330	GR House GR 房子
264	Garden 15 花园 15	334	JN House JN 房子
266	Going Green 走向绿色	340	AMB House AMB 房子
270	Leafy Entertainer 多叶表演者	344	Remanso de Las Condes, Casa C 雷曼索德拉斯康德斯，C 住宅
274	Kendall Residence 肯德尔住宅	350	San Carlos 1 圣卡洛斯 1
282	Glazer Residence 格莱泽之家	354	San Carlos 2 圣卡洛斯 2
288	Private Garden – Dalkey, Co. Dublin, Ireland 都柏林多基私家花园	358	House in Ramat Hasharon 拉马特·哈莎伦的房子
296	Roxbury Renovation 罗克斯伯里之家翻新	360	Narla 娜拉住宅
300	Mediterranean Garden 地中海花园	364	Fuleihan Residence Fuleihan 住宅
302	Garden of a villa in Dahlem 达勒姆花园别墅	370	Private Residence "Water Canvas" 私人住宅"水上帆船"
308	Squire Creek Residence 斯格尔溪住宅	374	Winter Residence 冬日住宅
314	House on the Hills 山上的房子		

Residential Community
居民小区

Larkspur Courts

拉克斯普庭院

Landscape Architect / SWA Group
Client / Lincoln Property Company
Location / Marin County, California, USA
Photographer / Tom Fox

The purpose of the project is to reclaim an abandoned quarry into an attractive residential village. The project required an interdisciplinary approach involving architects, civil engineers, geotechnical consultants and horticulturists.

Larkspur Courts is proving to be a successful reclamation project. The property has a very low vacancy rate. The leasing office often gets compliments from residents on how pleased they are to live there and how they love the landscape.

Our goal was to convert the abandoned rock quarry into an attractive new hillside residential community. To achieve this goal, we proceeded to employ the following concepts: Mass and site buildings to maximize offsite view potentials.

Create a pedestrian precinct within the development — emphasize the pedestrian connections and amenities. Restrict vehicles to the perimeter of the site. Create a distinct identity within each courtyard cluster.

Given the City's desire to create 97 family units, defined as two and three bedroom units no more than one story above grade, integrate these units with non-family units via building clustering and shared courtyards.

On a steeply terraced site, take up grade with building clusters and retaining wall. Generate usable outdoor spaces.

该案目的是将废弃的采石场变为一个有吸引力的住宅小区，需要一个跨学科的设计方法，需要建筑师、土木工程师、地质技术顾问和园艺师的协力合作。

Larkspur Courts 的设计证明了其是一个成功的改造项目。该地块具有非常低的空置率。租赁办公室经常受到居民的赞美，说他们住在那里是多么开心，是多么地喜爱那里的风景。

我们的目标是将废弃的采石场变成一个有吸引力的坡地新住宅社区，为了实现这一目标，我们采用了以下设计理念：聚集和定位建筑，最大化小区外的潜在风景。

我们计划在该地区创建一个行人专用区，强调行人关系和便利设施；限制周边地区的车辆；使每个庭院建筑群具有明显的特征。

考虑到市厅建造97户家庭住宅单位的要求，定义是不超过一层的两卧室和三卧室的住宅；通过建筑聚集和共用的庭院将这些家庭住宅和非家庭住宅融为一体。

在大坡度的阶梯区，继续分段建造建筑群和保留墙壁，形成可用的户外空间。

014 Residential Community 居民小区

Kingkey Oriental Grace

京基御景东方

Landscape Architect / PLACE Design Group Pty Ltd.
Client / Kingkey Group
Location / Shenzhen, China
Project Site / 50,000 m²
Photographer / PLACE Design Group

Kingkey Oriental is a mixed-use development delicately designed on a combined 'Form and Function' philosophy with the commercial precinct provided with a wide open frontage for outdoor retail functions, security and recreations for the kindergarten and residential precinct inspired by the need of relaxation and privacy.

Greatly dictated by the multi-level building podium each precinct was linked with escalators and stairs strategically placed within the garden with great views of the naturalistic water features.

京基御景东方是一个综合建筑项目，经过精心设计，结合"形式与功能"的理念，商业区为户外店铺提供了大量的正面空地，为幼儿园提供了安全和娱乐空间，为住宅区提供了放松和隐秘的环境。水景大多是被多层建筑裙楼支配，每个区域以自动扶梯和放置在美丽自然水景花园内的楼梯相连。

Residential Community 居民小区

Vanke Rancho Santa Fe II

万科兰乔圣菲二期

Landscape Architect / Guangzhou Homy Landscape Co., Ltd.
Client / Vanke Group
Location / Huadu, Guangzhou, China
Area / 41,000 m²

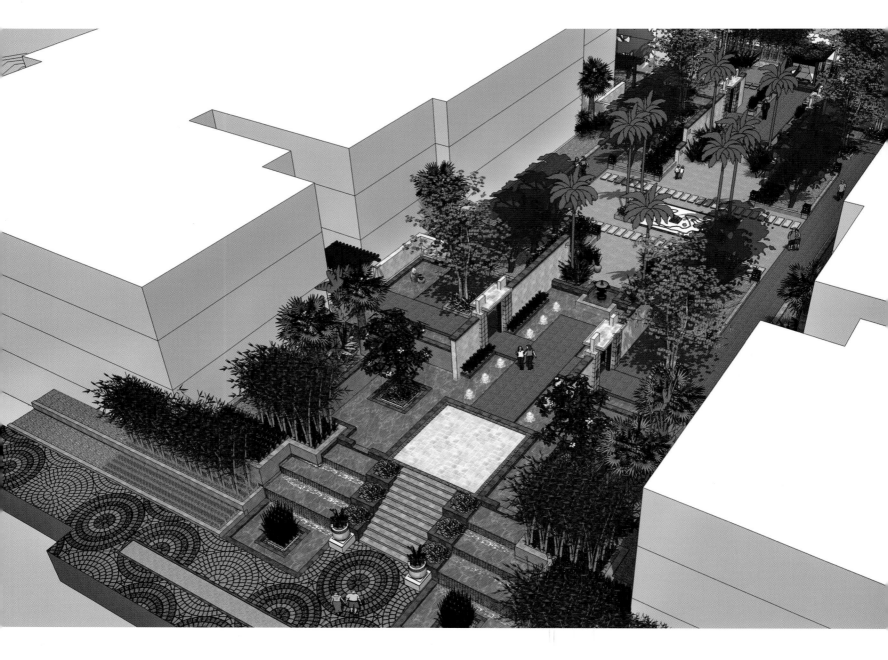

Rancho Santa Fe II applied modern Southeast Asian style. The main entrance is an annular driveway which both solves traffic problem and creates an open space. The wooden construction at the entrance and the water feature which combines the fountain with mirror pond in the center highlight the Southeast Asian style. It comes to the point and makes the focal, presenting distinguished and grand sense. The pure grass slop on both sides with terraced plantings brings people with fresh and special experience.

万科兰乔圣菲二期采用现代东南亚风格，主入口区采用环形车道，既解决了车流问题又营造了开阔的空间。入口木质构筑物和中心以喷水和镜面水池结合的水景突显东南亚风格，开门见山，突出重点，给人尊贵大气之感。两侧纯净的梯级草坡，配合层级的种植设计，带给人们清新、特别的感受。

Kingold HuiJing New Town – LongXi Mountain Park

侨鑫汇景新城—— 龙熹山山顶公园

Landscape Architect / Guangzhou Homy Landscape Co., Ltd.
Client / KINGOLD Group
Location / Guangzhou, China
Area / 70,000 m²

The intention of this project lies in creating top quality outdoor environment for the residential community in Southern China, which makes the resident distinguished yet without losing the cordial and friendly atmosphere. The design created a noble and yet natural space and finished with the continuous landscape axis highlights.

The design indicated environment friendly concept. Simple and typical forms expressing the environment-related intention were quoted to enhance the overall image, with additional functional spaces to meet users' practical needs. With collection and utilization of reclaimed water and rainwater, the ecological design concept was represented throughout the site. The eastern area has become the most natural and ecological part in Huijing New Town because of the golf course and mountain park.

本项目欲构建中国南方顶端居住社区的室外环境，让居住者产生尊贵又不失亲切、温馨的感受，营造高贵又不失自然的品位空间，并成为连续景观轴高潮的收尾。

该项目的设计力求表现绿色环保的主题，为追求整体形象引用了单纯、典型的形态语言，表达了环境亲和性意向，并增加各个功能空间以满足人们的实际需要。通过中水及雨水的收集与利用，使生态设计理念贯穿整个设计，并且由于高尔夫场地及山地公园，东区成为汇景新城最富自然生态气息的部分。

Times Times Peanut

时代 • 时代花生

Landscape Architect / Guangzhou Homy Landscape Co., Ltd.
Client / Guangzhou TIMES Group
Location / Guangzhou, China
Area / 27,000 m²

The development focuses on the subtropical landscape with modern elements. The floating lines and simple straights consist of the well-defined functional landscape space, which embodies the overall layout with rational as well as zest. The waterscape as landscape axis links up the east-west oriented water system and unifies the design throughout the site. When it comes to the design for central landscape space, contrasts from dynamic and static, open and private, modern and rural create a rhythmic space. Meanwhile the small water features on the overhead layer activate and enrich the space. The design of the emergency path applies smooth, lively and free curve linear construction and combines with the green lawn, which makes the site's walkways more interesting and in which residents can occupy and enjoy this comfortable space filled with modern aura.

本方案着重于表现亚热带风情园林景观，同时加入现代元素，以流畅的线条和简练的直线构图，组成功能明确的景观空间，使整体布局赋予理性，又包含激情。贯通东西向的水系景观作为景观轴，统一整个小区的设计。在中心景观空间的设计上，运用了动与静、开放与私密、现代与野趣的对比，创造出有节奏、有韵律的空间。同时在架空层设计了多处小型水景，使空间变得活跃和丰富。消防通道的设计采用流畅、活泼、自由的曲线构图，并与草坪绿化相结合，使小区道路变得丰富而有趣，居民可尽情地拥有和享受这个洋溢着现代气息的舒适空间。

036 Residential Community 居民小区

Zhuhai Nanfu Jinyuan Landscape

珠海南福锦园景观设计

Landscape Architect / L&A Design Group
Location / Zhuhai, China
Client / Nanfu Real Estate Development Co., Ltd.
Area / 33,674.38 m²

Zhuhai Nanfu Jinyuan landscape design project is the reconstruction project for Guancun Garden. It leans on Banzhang Mountain in the north, faces Jiuzhou Avenue in the south, borders on Bailiandong Garden in the east and looks onto the Australia Villa in the west. The location is richly endowed by the nature and full of landscape resources, which makes it the rare treasure site in the heart of Zhuhai downtown.

Design Concept

Modern – Modern Thai landscape design interprets the essence of the project planning, creating a human-oriented, green and classical resort.

Nature – Base on the natural landscape in Banzhang Moutain and Bailiandong Garden, the project introduces the untouched natural landscape into the resort which continues the green from the nature.

Healthy – It makes full use of the location advantages and converts Banzhang Mountain into a mountain park for community leisure activities, improving the sound and healthy quality in "Nanfu Jinyuan".

Design Principles

The project makes full use of the untouched natural landscape in Banzhang Moutain and Bailiandong Garden and analyzes the possibilities of landscape and space construction, creating a relaxing, pleasant landscape that has district gradations and meets the needs for communication, recreation, and entertainment.

i) Its symbiosis with the nature, that is, fully blends into the existing environmental resources in the surrounding areas, so as to retain the original environment and natural resources and enhance the whole greening and ecological scale, enabling the seamless combination of the landscape, nature and buildings.

ii) The design gives each functional component characters and identification through landscapes.

iii) The language of the design is modern and subtle, meeting the functional demands and is operational.

iv) The design aims at creating a lush, pleasant and freely accessible community.

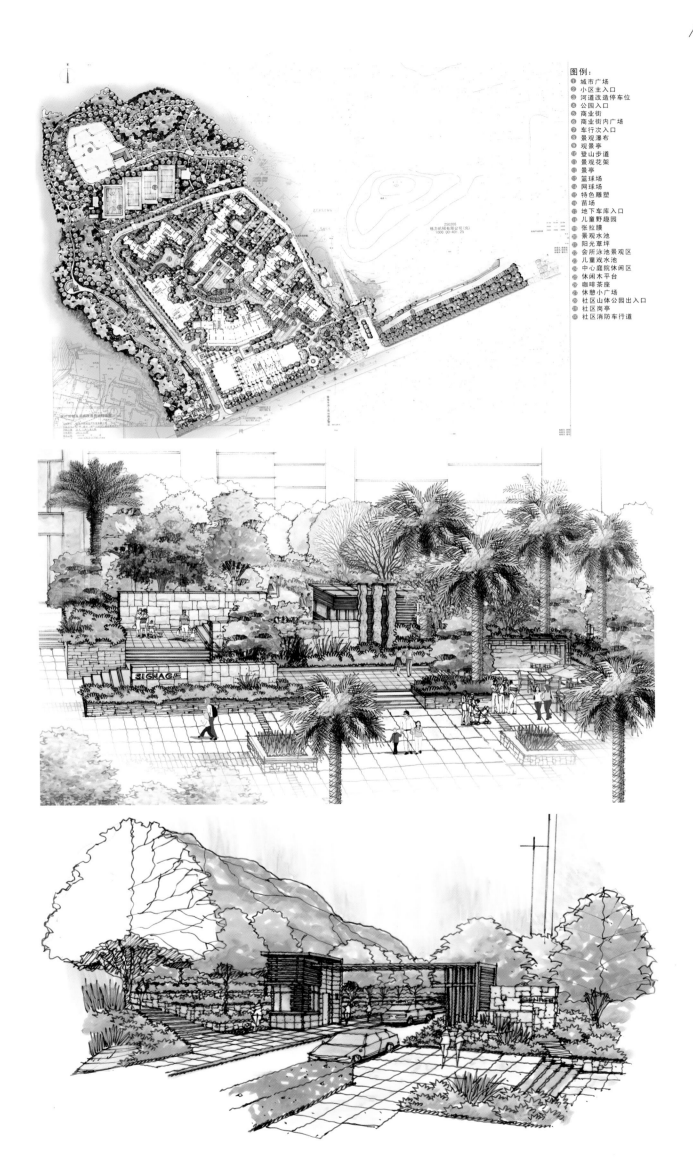

图例：
① 城市广场
② 小区主入口
③ 河道改造停车位
④ 公园入口
⑤ 商业街
⑥ 商业街内广场
⑦ 车行次入口
⑧ 景观瀑布
⑨ 观景亭
⑩ 登山步道
⑪ 景观花架
⑫ 景亭
⑬ 篮球场
⑭ 网球场
⑮ 特色雕塑
⑯ 苗场
⑰ 地下车库入口
⑱ 儿童野趣园
⑲ 张拉膜
⑳ 景观水池
㉑ 阳光草坪
㉒ 会所泳池景观区
㉓ 儿童戏水池
㉔ 中心庭院休闲区
㉕ 休闲木平台
㉖ 咖啡茶座
㉗ 休憩小广场
㉘ 社区山体公园出入口
㉙ 社区岗亭
㉚ 社区消防车行道

珠海南福锦园景观设计为官村花园改造项目，北靠板障山，南面九州大道，东邻白莲洞公园，西望澳洲山庄，地理位置得天独厚，景观资源优越，是珠海城市中心腹地不可多得的风水宝地。

设计理念

现代——以现代泰式的景观设计风格，诠释"南福锦园"项目规划的精髓，打造人性化、绿色、经典的度假社区。

自然——以板障山、白莲洞公园的自然景观为依托，将板障山和白莲洞的原生态自然景观引入社区，延续大自然的绿色。

健康——充分利用项目的地域优势，将板障山纳入社区休闲活动的山体公园，提高"南福锦园"阳光、健康的社区品质。

设计原则

充分利用板障山、白莲洞原生态的自然景观，分析环境可能形成的景观空间结构，营造一个轻松、愉悦、富有层次的景观来满足人们交流、休憩、娱乐等需求。

1、与自然共生，与现有周边地域环境资源充分结合，尽量保留原生态自然环境资源，提高整体的绿化率和生态化，使景观、自然、建筑完美地结合。

2、通过景观赋予每个组团功能特色与可识别性。

3、景观的设计语言现代、精致、满足功能并具可操作性。

4、创造一个葱绿、宜人、可自由通行的高品质社区。

Stanford West Apartments

斯坦福西区公寓

Landscape Architect / SWA Group
Client / Stanford University
Location / Palo Alto, California, USA
Photography / Tom Fox

Located in the Sand Hill Road corridor of the Stanford University Campus, Stanford West Senior Housing consists of 388 residential units for senior citizens together with 70 assisted living units and a 46 room skilled nursing facility. Located on the former site of a children's hospital, the design concentrates the building program in the middle of the site in order to preserve existing trees, provide increased setbacks from Sand Hill Road and the creek, and allow seniors to circulate comfortably around the facility. Stanford West apartments consist of 630 multifamily units and are oriented to employees at nearby employment centers such as the campus, medical center, and research park. The new neighborhood is currently planned to preserve natural features on the parcel, including archaeologically sensitive areas and existing mature trees. Internal street grids, architecture and landscape elements are designed to recall the traditions of existing older neighborhoods in Palo Alto and Menlo Park.

位于斯坦福大学校园沙丘路通道的斯坦福西区老年人之家包括为老年人建造的388套住宅单位，以及70套辅助住宅和46间专业护理室。该项目坐落在一家儿童医院旧址，为了保留现有树木和进一步阻隔沙丘路和小溪，以及让老年人可以舒服地在附近转悠，该设计侧重于中间的建筑项目。斯坦福西区公寓有630多户住宅单位，是给在附近工作的员工提供的，如在校园、医疗中心、研究公园上班的员工。新社区目前计划保留该地的自然特征，包括考古区和现有的高大树木。内部街道、建筑和景观的设计让人想起在帕罗奥图和门罗公园现存老城区的传统。

/ 043

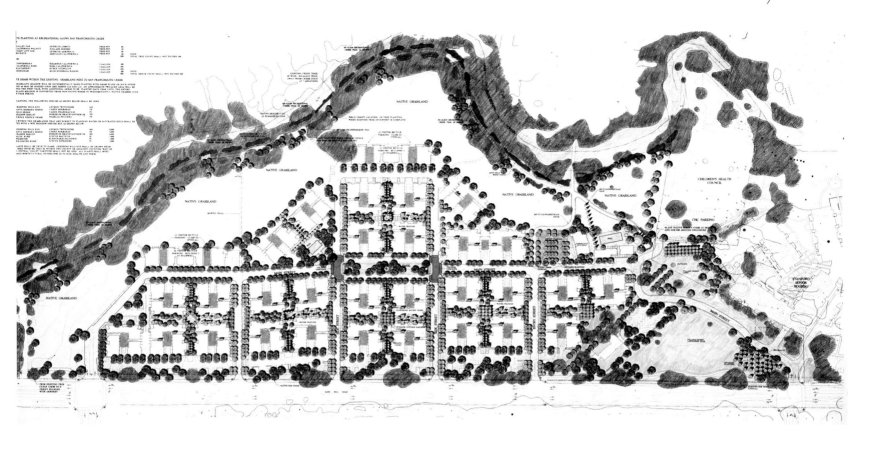

Hangzhou - Spring River Flower Moon

杭州——春江花月

Landscape Architect / PLACE Design Group Pty Ltd.
Client / Zhenjiang Greentown Real Estate Group
Location / Hangzhou, China
Area / 150,000 m²
Photographer / PLACE Design Group Pty Ltd.

Water is used throughout the site in a variety of ways to fulfill a variety of functions. The landscape masterplan is divided into four distinct segments by the two major axes which intersect like lines of latitude and longitude across the site. One axis runs laterally from the south western entry point to the lake. This axis is interrupted by the extensive planting of trees on the far side of the lake. This is intended to provide a soft green termination to the view as you progress along this axis. The other axis runs from the clubhouse to the main road that forms the southeastern boundary. The views down this axis are designed to focus on the river to the south east and take in the main lake and its surrounding landscape.

The main lake is designed at the intersection of the two major axes and provides the focal point for the entire site. This central parkland is further divided by radiating axes designed to direct and reinforce views from the clubhouse towards the river and backwards towards the clubhouse.

By using contrasting landscape techniques, varying between formal, semi-formal and natural, it is intended that the total design will provide the residents and users with a very high quality landscape facility that better responds to the human desire for landscape beauty (an ordered disorder) and, in particular, focuses with sensitivity upon the social, aesthetic, cultural and recreational aspirations of the intended user.

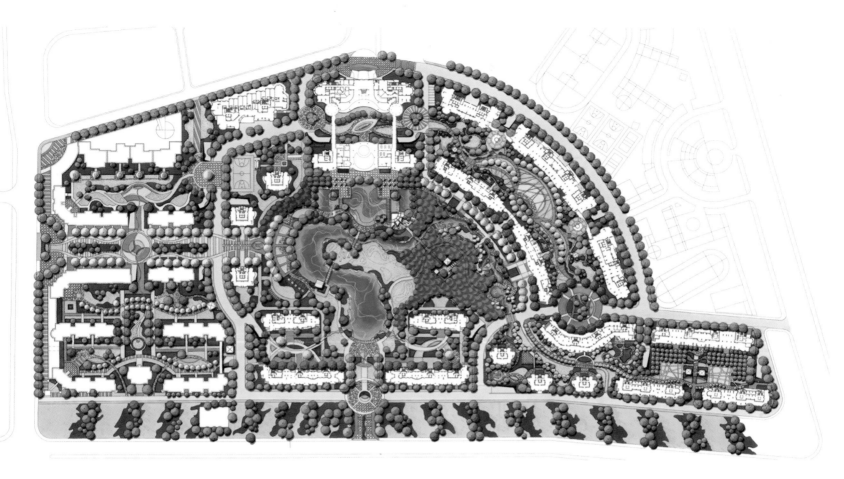

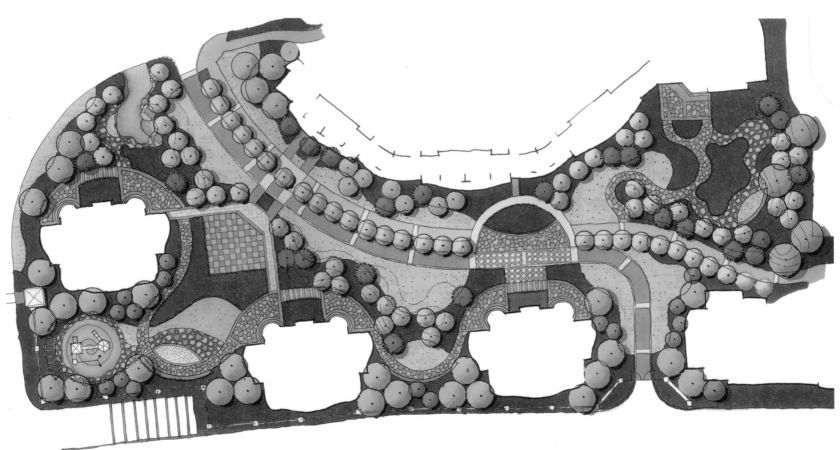

水以各种方式使用在整个小区以满足不同的功能。景观总规划通过两条主轴线划分为四个不同区域，主轴像经纬线一样在小区相交。一条轴线从南西入口处横穿到湖边，这条轴线被湖的另外一边的树木中断，目的是让行人停下来观赏柔和的绿色风景。另一条轴线从会所穿到界定东南边界的主道上，此轴线南边的景色设计侧重于东南边的水域，方便居民欣赏主湖及其周边景观。

主湖设置在两条轴线交汇处，是整个小区的中心焦点。中央绿地由发散的轴线进一步分区，轴线设计是指示和增强从会所到河流的景色。

通过对比形式与半形式和大自然之间的不同景观，该设计旨在为居民和用户提供优质的景观设施，更好地反映人类对景观美（一种有序的混乱）的渴望，尤其是侧重于预期用户的社会、美观、文化和娱乐需求。

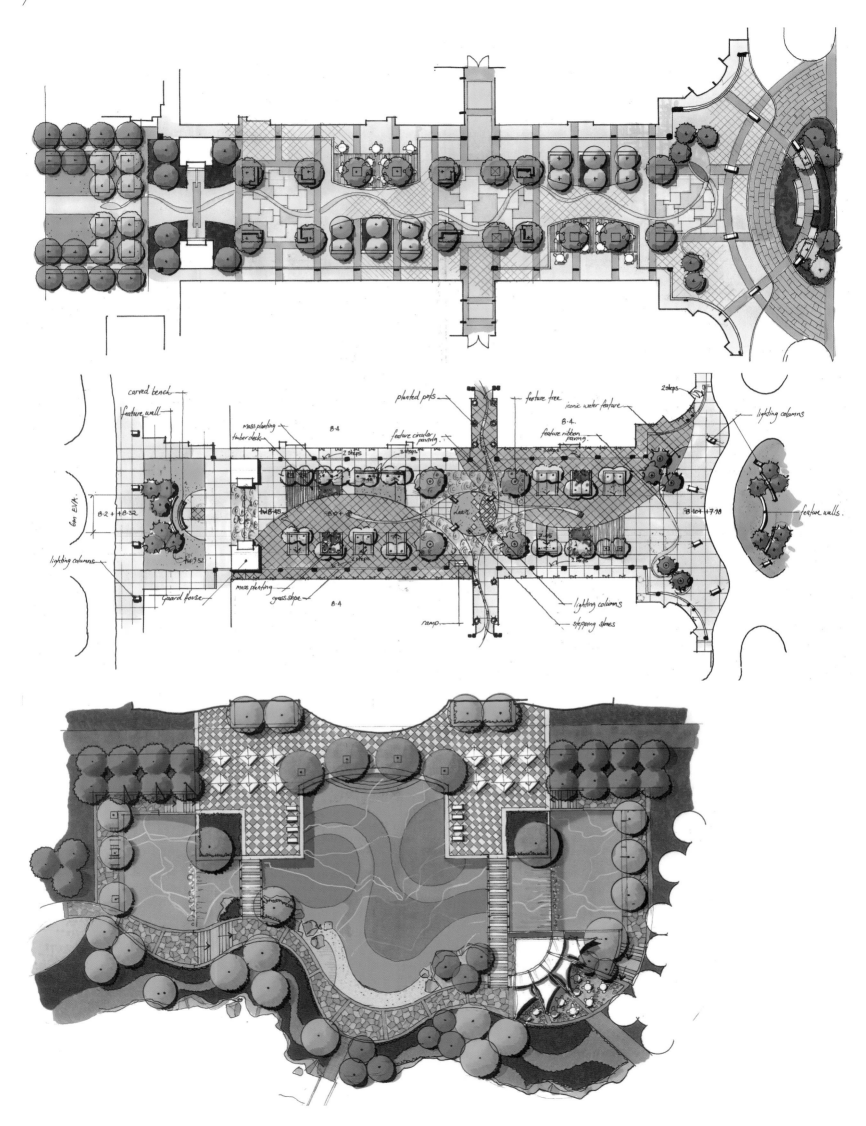

Dakota Residences

达科塔住宅

Landscape Architect / ONG&ONG Pte Ltd.
Team Director / Lena Quek
Client / Rivershore Pte Ltd.
Location / Singapore
Photographs / Courtesy of ONG&ONG Pte Ltd.

Ideally located near two major highways, this is a highly accessible development with stunning views of the nearby Geylang River.

The condominium offers a river-side lifestyle, seen in the sloping water feature at the forecourt, and the lush greenery that swathes its landscape. It also feels like a spa oasis, with a mineral spa pool that provides users with water massage jets, spa beds and seats. A 50-metre lap pool, attached Aqua gym, lounge pool, as well as a string of outdoor fitness stations cater to all exercise enthusiasts.

Children can have their fun as well, with a children's splash zone, and play ground located on the top tier of a unique 2-tier garden. The lower tier also has BBQ pits where the adults can relax, while still being able to keep watch of their children.

该案位于两条主要高速公路附近，是一个交通四通八达的理想位置，并且可看到附近芽笼河的美景。

该复式公寓提供了一种河畔生活方式。从前院的斜坡水景和围绕景观的大片葱郁的绿色，看起来感觉像一个绿洲；加上为用户提供喷水按摩、SPA床椅的矿物温泉池，一个50米长的游泳池与水式健身房相连，以及一系列的户外健身站满足了所有运动爱好者的需求。

孩子们也有他们的玩乐空间，儿童飞溅区和娱乐场所坐落在独特的二层花园的顶层。下层也有烧烤坑，在那里成年人可以放松休闲，同时还能够照看他们的孩子。

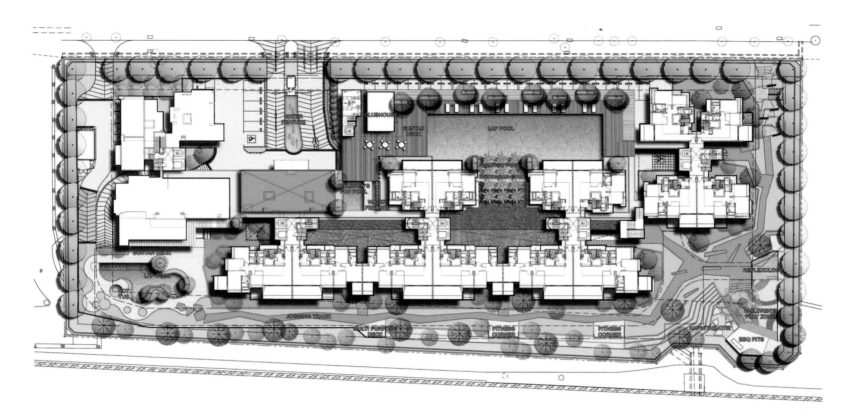

/ 064 Residential Community 居民小区

Zobon City Sculpture Garden

中邦城市雕塑花园

Landscape Architect / SWA Group
Location / Shanghai, China
Client / Landmark Development Consulting (Shanghai) Co., Ltd.
Area / 20,000 m²
Photographer / Tom Fox

The Zobon City Sculpture Garden lies at the center of a 5,000 unit multi-family residential infill development in the Pudong district of Shanghai. The design objective was to create an innovative model for multi-family habitation that integrates art, landscape and architecture in ways that make dense, urban living more sustainable. On a mere 0.6 hectare site, the landscape architecture expresses three moments which celebrate the inherent, and often invisible, beauty of the city:

The Huang-Pu Abstraction garden mimics the seasonal flooding along the banks of the Pu River by creating a 2.5 meter tall stacked, glass fountain (flood level) that cascades water along its length. A public plaza provides informal seating and a place to view the water feature.

The Cognitive Garden is designed to offer residents and visitors a place of respite for meditation and relaxation and is infused with colorful planting, raked gravel, and white sound from a simple fountain.

The Sky Garden is the centerpiece of the project. It attempts to capture what is left of the diminishing skyline by reflecting it directly on to itself through an elliptical-shaped reflecting pond filled with Koi fish and ringed by native plantings and aeration terraces. It is also a gesture towards Shanghai's freeway infrastructure and land subsidence through its curving forms and elevation changes. The outdoor program for this area includes places for tai-chi, eating, small gatherings and walking/viewing.

中邦城市雕塑花园位于上海浦东区一个 5000 户家庭住宅单位的开发区中心。设计目标是为多户居住打造一种创新模式，集艺术、景观和建筑于一体，让密集的都市生活方式更有可持续性。在仅 0.6 公顷的土地上，该景观建筑表达了三个时刻，这三个时刻赞美了内在的且往往是隐形的城市之美。

黄浦抽象花园沿着黄浦江畔模仿季节性洪水，打造一个 2.5 米高的堆叠玻璃喷泉（洪水位），水流倾泻而下，公共广场提供了方便观看水景的休闲式座位。

认知花园的设计是为了给居民和游客提供冥想和放松的地方，这里布满了丰富多彩的植物，倾斜的砾石路和来自简单喷泉的白色水流。

空中花园是该项目的核心，它试图捕捉逐渐缩小的天际线所剩下的空间，通过直接反射，到达一个椭圆形的、被原生植被和通风台阶圈起来的鲤鱼池。这也是一种对上海的快车道基础设施和地面沉降的表现。这个地区的户外项目包括太极、餐饮、小聚会和步行观景道。

Mantra Aqua Resort

Mantra Aqua 度假村

Landscape Architect / PLACE Design Group Pty Ltd.
Client / Project Control
Location / New South Wales, Australia
Photographer / PLACE Design Group

Significant stands of mature trees endemic to the area were integrated into the layout of this project nestled high in the woodlands overlooking Nelsons Bay.

A strong elliptical pool form was incorporated with special emphasis on "aqua-play" and outdoor dining. Previously an abandoned development site, the project has provided the catalyst for further sustainable development in the Central Coast region.

该项目坐落在林地，俯瞰着纳尔逊湾，该地区特有的高大树木被融入到项目的布局中。

一个极大的椭圆形池被合并到特别强调的"嬉水场"和户外用餐区。该案之前是一个废弃的地方，它为将来中央海岸区的可持续发展提供了催化剂。

Circle on Cavill

卡维尔之圈

Landscape Architect / PLACE Design Group Pty Ltd.
Client / Sunland
Location / Gold Coast, Australia
Photographer / PLACE Design Group Pty Ltd.

Circle is located at the heart of the Gold Coasts re-born urban precinct. The visually stunning high rise residential development is an award winning example of mixed use retail spaces at street level. Circle on Cavill has quickly become the new benchmark for Surfers Paradise urban design.

Water feature placement and plant selection have been made with careful consideration of winds, view lines and amenity of adjacent residences.

Offering heated, lap and lagoon style pools overlooking the retail precinct, Circle on Cavill recognizes the residents need for relaxation away from the 'buzz' of Surfers Paradise. The landscape scale has been sensitive to the building resident's requirements, integrating social zones; children's play areas, with BBQ facilities to allow for different levels of interaction and personal privacy.

The iconic architecture of Circle on Cavill stands out against the Surfers Paradise skyline and serves as a readily identifiable place marker from afar. The success of Circle on Cavill exhibits an understanding of the critical synergies, that combined are the three triple bottom line elements of sustainability: social, economical and environmental.

　　Circle 位于黄金海岸重建城区的心脏地带，这栋外观宏伟的高层住宅大楼是一个街面零售空间综合体的获奖案例。"卡维尔之圈"已经迅速成为冲浪者的天堂、城市设计的新基准。

　　水景布局和植物选择都仔细考虑了风向、景观线路和相邻住宅的舒适性。

　　圆形泻湖式的可加热游泳池俯瞰着零售区，"卡维尔之圈"认识到居民远离"噪杂的"冲浪者天堂的放松需求。景观规模与社会环境相结合，根据居民要求建造；儿童游乐区，以及烧烤设施促进了不同层次的互动同时也保证了个人隐私。

　　"卡维尔之圈"的标志性建筑矗立在冲浪者天堂的天际线中，且作为从远处都易于识别的标记。该项目的成功显示了对关键的协同效应的理解，这种结合是三种基本要素的可持续性：社会、经济和环境。

The Trillium

延龄草社区

Landscape Architect / ONG&ONG Pte Ltd.
Team Director / Lena Quek
Client / Lippo Land Corporation
Location / Singapore
Photographs / Courtesy of ONG&ONG Pte Ltd.

At the entrance, a majestic water feature greets you while a string of reflective pools continue round the estate. Residents may lounge in shaded comfort at the main pool, which branches off into a children's pool where little ones can play as well.

A unique feature is that the pathways are lit by fire display columns, which are swathed in flora to give visual relief to the landscape. The site's peripheral is also lavishly bathed in greenery for privacy as well as aesthetic delight.

With its urban design and green landscape, The Trillium is the perfect marriage between Man and Nature.

入口处，一串反射池围着建筑的同时，一个壮丽的水景迎接你的到来。居民可以舒服地躺在主池的荫凉处，主池中有儿童游泳池，小朋友也可以尽情地玩乐。

该案的一个特色是呈火柱排列的通道，火柱围绕在植物中，为景观提供了视觉变换。该建筑的外围也沐浴在葱翠绿色中，不仅满足隐私方面的要求，也体现了审美情趣。

城市设计和绿色景观中的延龄草是人与自然之间的完美结合。

Bamboo Garden

竹园

Landscape Architect / PLACE Design Group Pty Ltd.
Client / Jincheng Group
Location / Hangzhou, China
Photographer / PLACE Design Group Pty Ltd.

PLACE Design Group started work on this project in 2004. PLACE has provided design services for all stages of the works.

Extending over approx. 39 ha Bamboo Gardens contains a mix of high rise, low rise and villa apartments set within extensive community recreation spaces.

PLACE 设计集团在 2004 年开始这个项目，负责该项目所有阶段的设计。

在宽敞的游憩社区空间里，绵延约 39 公顷的竹园中包括了高层、低层和别墅公寓。

Residential Community 居民小区

Shanghai Greentown

上海绿城

Landscape Architect / PLACE Design Group Pty Ltd.
Client / Greentown Group
Location / Shanghai, China
Photography / PLACE Design Group Pty Ltd.

The overall design philosophy seeks to give the development the feeling of being a forest where there are places to live, places to gather and play and quiet places to wander and reflect. The 'forest' within the site reinforces the feeling of being surrounded by vegetation as one would be in a forest, following a pathway to another clearing where there is something else to discover.

Eight groups of residential high rises make up this development ringed around a central community focal point – the leisure centre. The central leisure centre contains restaurants, swimming pool and playground facilities offering numerous amenities for residents. Each group of buildings is separated from others by groves of trees, while the central ring road surrounds the Leisure Centre and provides a link between different areas of the development.

Each residential area of the site has its' own recreation facilities, open active spaces and passive reflective areas, including water features, amphitheatre areas and children's playgrounds. The clever use of land mounding and vegetation creates interest and a sense of discovery throughout the project.

The success of this design is in the sense of oasis as soon as one enters the site. The wide areas of dense layered planting effectively buffer the noise and pace of the surrounding city and allow the user to believe they have been transported far away back in time even to the more rural past of this Pudong area. The desire of the Client to create a forest in the city has been realized in the construction of this development.

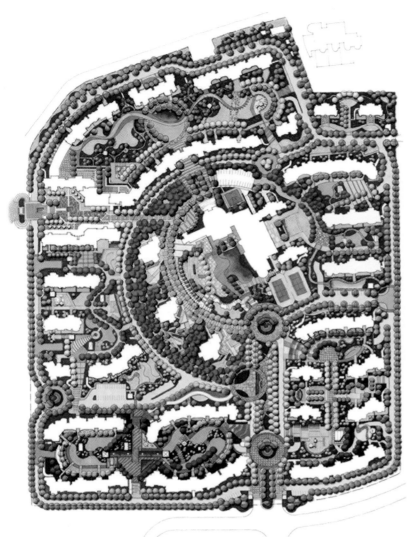

该案整体设计理念旨在体现森林般的感觉，在那里有居住、聚会和玩耍的地方，以及一个可以漫步和冥思的静谧之地。该地的"森林"增强了被植物包围的感觉，就好像真的是在森林里一样，随着一条小径走到另一条可以发现些什么的林中空地。

八组住宅高楼构成了这个环绕着一个中心社区的焦点——休闲中心的项目。中央休闲中心包括餐厅、游泳池和为居民提供众多娱乐活动的游乐场。每一组建筑群由小树林跟其他组分开，而中央环道则环绕着休闲中心，成为不同区域之间的纽带。

每个住宅小区都有自己的娱乐设施、开放的活动空间和其他空间，包括水景、圆形露台和儿童游乐场。土丘和植物的巧妙运用表现了趣味性和处处存在的发现感。

本设计的成功之处是一走进这里就有进入绿洲的感觉。密集的大面积种植有效地减轻了周边的都市噪音和速度，让业主相信他们已经远离城市，甚至是回到过去浦东区的农村之中。

Yanlord Riverside Home St III

仁恒河滨花园第三期

Landscape Architect / PLACE Design Group Pty Ltd.
Client / Shanghai Yanlord Property Co., Ltd.
Location / ShangHai, China
Area / 85,000 m²
Photographer / PLACE Design Group Pty Ltd.

Yanlord Riverside Development is the 3rd stage of a 25Ha residential development in Shanghai, China. Place was asked to design and document this 8.5 Ha site including Vehicular & Pedestrian Circulation, Community Centre Facilities, Children & Adult Pools, Central Water Feature, Riverside Trail, 300+Ampitheatre, Tennis Courts, BBQ facilities, children's playgrounds & Courtyards. There are 12, 20 storey buildings accommodating approx. 1500 residents more challenging aspects of the project include integration with other Development Stages, The landscape on podium restricts internal access roads.

仁恒河滨花园开发项目是上海一个 25 公顷的住宅项目的第三期。PLACE 设计集团负责设计和规划 8.5 公顷的地段，设计内容包括车辆和人行通道、社区中心设施、儿童与成人游泳池、中央水景观、河滨小径、直径 300 米的圆形露天剧场、网球场、烧烤场、儿童游乐场和庭院。有 12 栋 20 层的建筑，共可容纳约 1500 个居民。更具挑战性的方面是，该案也融入到了前两期中，裙楼的景观为小区的内部通道提供了屏障。

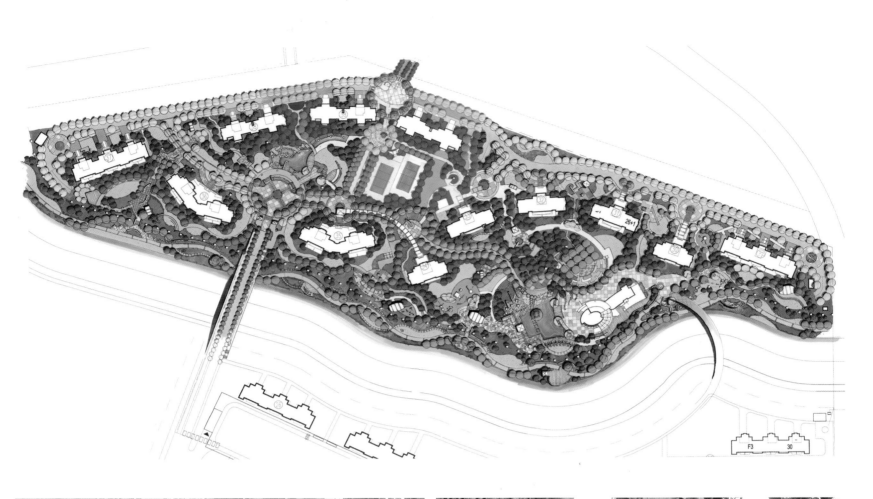

Riverlight Apartments

Riverlight 公寓

Landscape Architect / PLACE Design Group Pty Ltd.
Client / Cape Bouvard
Location / Noosa, Sunshine Coast, Australia
Photographer / PLACE Design Group Pty Ltd.

The Riverlight apartments are located on an escarpment overlooking Noosa Sound and Laguna Bay in the Noosa tourist Mecca on the Sunshine Coast. This luxurious complex incorporates a generous resort-style outdoor recreation area, including a PLACE designed 25m lap pool, a child friendly lagoon pool and expansive gardens. Shelters, therapeutic spas, fully equipped gymnasium, yoga deck, barbeque and bar complete the communal facilities. The verdant sub-tropical vegetation integrates the resort with the surrounding natural coastal woodland environment.

The careful selection and design of landscape elements such as paving, fencing and water features compliment the architectural structures creating an overall cohesive design. The impressively landscaped grounds provide a resort-style setting.

Riverlight 公寓位于悬崖上，俯瞰着阳光海岸努沙旅游胜地的努沙湾和拉古纳湾。这豪华的综合体有一个广阔的度假风格户外游憩区，包括 PLACE 设计集团设计的 25 米游泳池、儿童专用泻湖游泳池和宽敞的花园。遮蔽处、水疗中心、设施齐全的健身房、瑜伽板、烧烤和酒吧组成完整的公共设施。青翠的亚热带植物将度假区与周边自然海岸林地环境融为一体。

精挑细选和精心设计的景观元素，如铺面、篱笆和水景跟建筑结构相符，创建整体连续的设计。令人印象深刻的地面提供了度假风格设置。

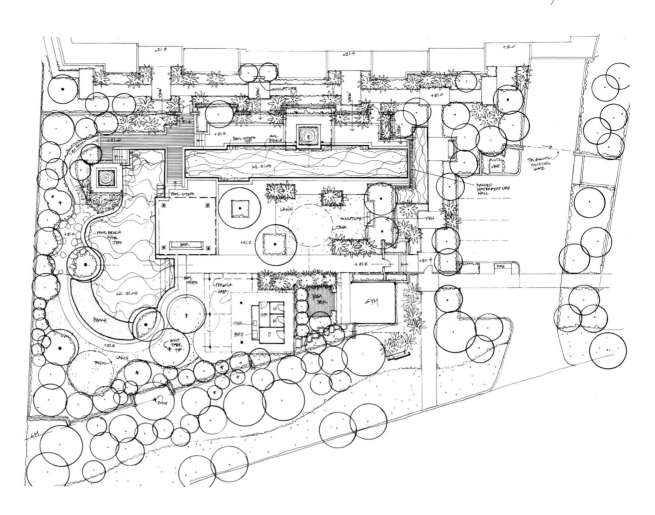

Oak Bay Stage II-III

橡树湾 II-III 期

Landscape Architect / PLACE Design Group Pty Ltd.
Client / CR Group
Photographer / PLACE Design Group Pty Ltd.

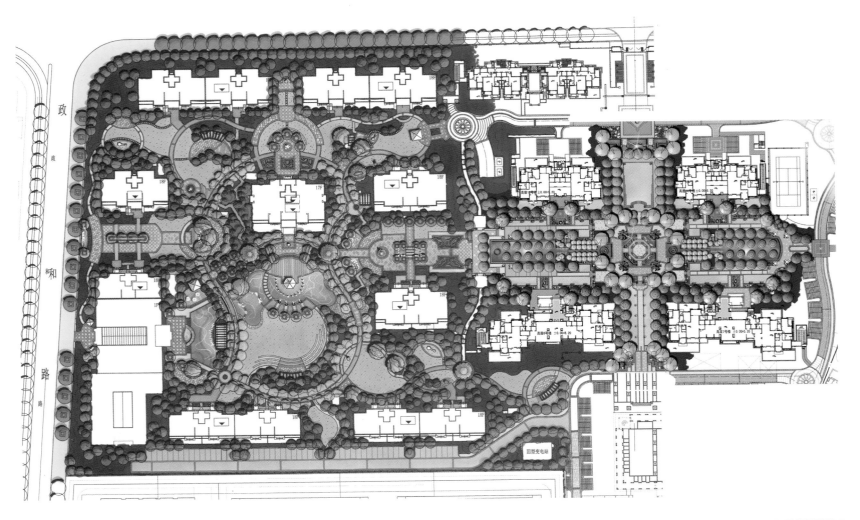

The Oak Bay Stage 3 briefly called for a natural style garden to create links to earlier stages of work (by others,) via a formal Art Deco inspired water axis. This axis has been carefully planned to allow for the seamless transition between different precincts of the overall project.

橡树湾第三期要求通过艺术装饰的水轴打造出自然花园风格，以延续前两期的风格（由其他公司设计的）。精心设计的水轴确保了整个项目中不同区域之间的完美过渡。

Shanghai Greentown Rose Garden

上海绿城玫瑰园

Landscape Architect / SWA Group
Client / Shanghai Greentown Forest Golf Villas Developing Co. Ltd.
Location / Shanghai, China
Area / 850,000 m²
Photographer / Tom Fox

The project has 239 single family estate and luxury lots, all South-oriented units and many with water view frontage. The design concept was to create a sense of place reminiscent of the "Old Shanghai Villas" built in the 1920s. Residences have a European ambiance within a forested environment. Privacy and security from adjacent lots and public zones is insured through use of high perimeter berms and walls.

The main idea is to create a sense of a very special place that is unique, exclusively high-end residential development that harmonizes both architecture and landscape and unites both European flavor as well as "Old Shanghai Villas" with integrated systems that residents will relate and enjoy as a comfortable, special place to reside. Architecturally, all the individual high-end single family villas have basically two lot sizes – luxury (1300m2+) and estate (4,000m2+). Building styles are reminiscent of traditional European luxury villas. The main landscape site planning idea is to provide a natural forested environment and incorporate waterways and together with the circulation system provide a unique private community with a system where public and private areas are related and cohesive, yet separated to insure privacy of the residents.

The clubhouse site is approximately 2.7 ha. (5.8 ac) and the site is the visual focus from the main south entrance. There is an existing historical church structure built in 1948 that will be retained in place and restored.

Clubhouse Landscape:

There are three main approaches for the landscape design.

1. Integrate grading and incorporating the natural overall forested environment to relate to the overall project image and insure both pedestrian and vehicular circulation fit the overall patterns.

2. Relate to the formal nature and materials of the clubhouse architecture utilizing the paving patterns, wall systems, fountain features, parterre gardens, clipped hedges, and site furniture, planter pots, and sculptures to relate to the architectural style.

3. The Main Areas of Landscape are: the east main arrival zone, the west courtyard zone, the south water oriented terraces, and the north interface zone to existing restored church structure and natural water edge.

该项目有 239 户独栋的庄园和豪华住宅，均是坐北朝南且许多住宅的正面拥有水景。设计理念是创建一个能让人想起建于 20 世纪 20 年代的"老上海的别墅"的地方。郁郁葱葱的森林环境中的住宅拥有欧式风情，通过高墙和堤壁来确保毗邻地段与公共区域的隐私和安全。

该案的主要设计理念是创建一个非常独特的空间，即独一无二、专享的高档住宅，协调建筑与景观，同时将欧式风情以及"老上海别墅"融为一体，是一个居民可以享受舒适、独特居住环境的地方。从建筑方面来说，所有高端独栋别墅有两种面积——豪华型（1300 平方米以上）和庄园型（4000 平方米以上），建筑风格是传统欧式别墅风格。主要景观设计理念是形成自然的森林环境和纳入水道，连同循环系统提供独特的私人社区，以及一个系统，在这个系统中，公共和私人区域既是息息相关的，也是分开的，以确保居民的隐私。

会所面积约 2.7 公顷（5.8 英亩），是南门入口处的焦点建筑。该处一座建于 1648 年的具有历史意义的教堂将会被保留和重新修葺。

会所景观

会所景观采用了三种设计手法。

1、将分级和结合自然整体森林环境融入到相关的整体项目中，保证行人和车辆的道路与整体模式相符。

2、将会所建筑采用的铺装图案、墙壁、喷泉水景、花坛、修剪整齐的树篱、家具、花盆和雕塑的材料与相关建筑风格结合起来。

3、景观的主要区域是：东边正门到达区、西边庭院区、南边水景露台区以及跟修葺的教堂和天然湖边相邻北边分界区。

Long Beach Villas

长堤花园别墅

Landscape Architect / PLACE Design Group Pty Ltd.
Client / Chang'an Company Group
Location / Shanghai, China
Photographer / PLACE Design Group Pty Ltd.

Taking residential villa development to a new level was the goal of our client on this project.

PLACE Design Group achieved this through creative freedom, transforming what would normally be left as road reserve into highly engaging public spaces.

将别墅住宅提升到一个新水平是我们客户对本案的目标。

PLACE 设计集团通过创造性的自主权，将通常意义上左边作为道路储备转变成了高度参与的公共空间，从而实现了这个目标。

Toscana

托斯卡纳社区

Landscape Architect / PLACE Design Group Pty Ltd.
Location / Changsha, China
Area / 210,000 m²
Photographer / PLACE Design Group Pty Ltd.

A multi-faceted residential and retail development features an extensive informal water spine bordered on both sides by formal and informal landscaped gardens. Tuscan and Californian architectural styles were translated into both structures and detail elements throughout the landscape.

该项目是一个多元化的住宅和商铺建筑，两侧有大量与园林花园相连的休闲水景。托斯卡纳式和加州式建筑风格通过景观融入到建筑的结构和细部中。

Vanke Rancho Santa Fe

万科兰乔圣菲

Landscape Architect / Guangzhou Homy Landscape Co., Ltd.
Location / Huadu, Guangzhou, China
Area / 38,000 m²

Rancho Santa Fe, an attractive place with dramatic hand-made plastering LOGO wall, hand-polished culture stones, and hand-made paintings, delicately reveals the native sense of California villa landscape. The pantile roofs in uneven levels used sophisticated materials. The log flowerpots add natural flavor to the buildings. The colorful curves that are made of rough iron wires, the high arch, the pots which are full of flowers…all these have absorbed the essence of Californian villa style, outlining the rich Californian style aura.

The landscape was designed to adhere with the noble and elegant Californian style villas. The natural and simple vertical entrance, native landscapes in the center, leisure and pleasant riparian views, constitute a landscape painting. To coordinate with the landscape, the design added some components such as pavilions and corridors, etc. The elements of the massive landscape buildings, small windows, plastering, solid walls, logs, etc. appear rough yet natural. The application of iron crafts presents simple and straightforward fashion. The exotic flavor reveals a mysterious feeling which fully respects the original texture and environment of the landscape, reflecting the local cultural heritage and landscape connotation.

兰乔圣菲，一个令人神往的地方，特色LOGO手工抹墙、手工打磨的文化石、徒手墙面抹灰，细腻、精致的细节透露出加州别墅景观构筑的原生感。高低错落的筒瓦屋顶用料考究，原木花架丰富了建筑的自然韵味。粗犷铁艺缠绵成缤纷绚烂的曲线，高高的圆弧拱门，种满花儿的陶罐……汲取了加州原汁原味的别墅风情，勾勒出浓郁的加州风情。

园景配合高贵、典雅的加州别墅风格建筑，古朴自然的纵深入口、自然生态的中心景区、休闲舒适的滨水风景区，组合成一幅风景画。园景配合建筑增加了景观亭、构架长廊等，景观构筑的厚实形体、小窗洞、抹灰、实墙、原木等特殊的材质显得粗犷、自然，强调铁艺的运用，呈现简洁、粗犷的时尚，体现一派异域风情，透露出一种神秘感，一种充分尊重原生地表肌理和环境的景观构筑，充分体现出当地的文化积淀与建筑景观内涵。

Conghua Hot Spring Villas

从化温泉别墅

Landscape Architect / PLACE Design Group Pty Ltd
Client / Guangzhou DEHE Investment & Development Co.
Location / Conghua, Guangdong, China
Area / 150,000 m²
Photographer / MF Advertising Agency

Conghua Hot Springs Villa is located in an area quite famous for its spring water. The PLACE Design Group worked closely with the clients to create stunning living environments that are fully coordinated with and complement architectural design. This project combines the unique advantage of geography of Conghua with the design of modern residential building to generate a high-quality residential area including its own Spas with natural spring water. The refreshing warm water in the Conghua Hot Spring froths to the surface at 12 different springs and the clear water are enriched with more than 10 kinds of rich minerals such as calcium, magnesium, and sodium and it enjoys an average temperature of 60 degrees centigrade.

The landscape design has been integrated with the modern and contemporary architectural design to create a very open and current atmosphere. The landscape design of the project uses introduction technique, which will take you from a scenic hillside to a water feature area at the entrance through to the community plaza where a large level change has been utilized in an imaginative way.

从化温泉别墅位于一个以温泉闻名的地区。PLACE 设计集团与客户密切合作，创造了与建筑设计充分相符和互补的令人惊叹不已的生活环境。本案结合从化的独特地理优势与现代居住建筑设计形成一个高品质的住宅区，包括天然的温泉 SPA。从化温泉别墅中令人神清气爽的温泉水，咕噜咕噜冒着泡，12 个不同温泉的水中富含 10 余种丰富的矿物质，如钙、镁、钠等，且平均温度为 60 摄氏度。

景观设计融入到现当代建筑设计中，创造了一个非常开放和流行的氛围。该案的景观设计采用引进技术，将带您从风景优美的山坡到入口处的水景区，穿过社区广场，那里有一处大型的层次变化，也采用了一种富有想象力的设计手法。

Magee Ranch

麦琪大农场

Landscape Architect / SWA Group
Client / The Broadmore Group
Location / Danville, California, USA
Area / 2,430,000 m²
Photographer / Tom Fox

The project site is a beautiful 540-acre hillside ranch in suburban Contra Costa County, California. It accommodates 257 custom and semi-custom homes. Over 1,000 oak trees were preserved by this design approach. When grading for the homesites was completed, the site retained its previous landform and tree cover. New landscape accentuated the natural beauty of the site by the use of stone, orchards and gardens. The entry road was designed to fit the hills and to create the mystery of a hidden valley where the homes were placed. At the entry, the magnificent oaks and creek were preserved as the entry feature.

The landscape architect's major contribution was to understand the dynamics of a complex tree-covered hillside site and the process of sensitively locating 300 homes without changing the basic landform and preserving the oak-studded hills. The landscape architect was responsible for site analysis, conceptual studies, land planning, detailed grading plans, and extensive public review through the City's approval process and community workshops. Following the City's approval of the design approach, the landscape architects provided design, implementation, and field observation of landscape for all common areas, streets, model homes, entry walls, bridges and open space areas.

This project was unanimously endorsed by all of the surrounding homeowner groups and received an unprecedented unanimous approval by the City of Danville Planning Commission and City Council. The mayor cited the unique accomplishment of preserving almost all of the 1,000 oaks on the property. The environmentally aware suburban community of San Francisco has repeatedly used this project as an example of how responsible development can be accomplished by sensitive land planning.

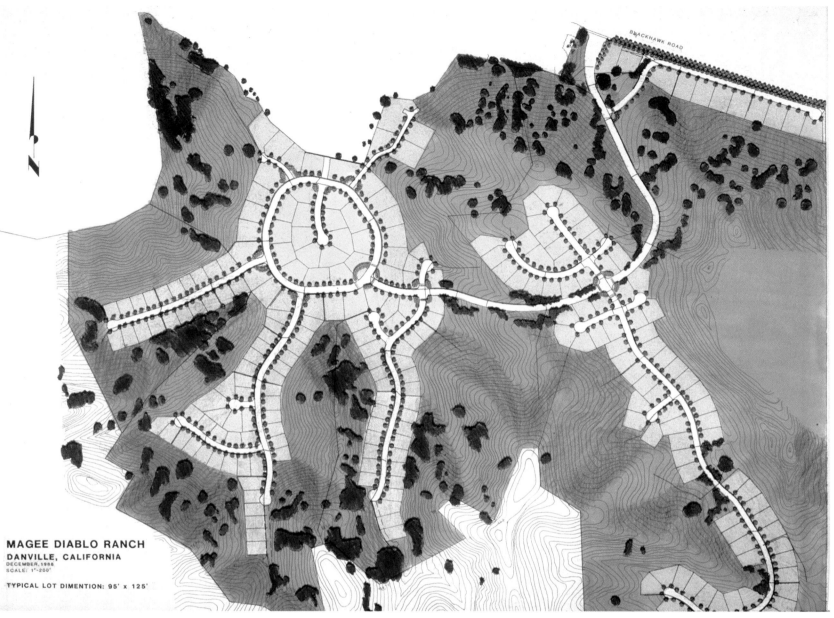

MAGEE DIABLO RANCH
DANVILLE, CALIFORNIA
DECEMBER, 1988
SCALE: 1"=200'

TYPICAL LOT DIMENTION: 95' x 125'

该案是位于加州康特拉科斯塔县郊区的一座 540 英亩的美丽山坡牧场，拥有 257 间定制和半定制的住宅。设计保留了 1000 多棵橡树，当房屋位置分段完成后，此地保留了其以前的地形和树木。新景观设计通过采用石头、果园和花园突显该地区的自然美。入口道路的设计是为了适应丘陵和创造隐蔽山谷的神秘性，房屋就设置在山谷中。入口处，华丽的橡树和小溪被保留作为入口处的特色。

景观设计师的主要作用是理解复杂的树木覆盖的山坡地块和在不改变基本

地貌和保留橡木林的前提下，灵敏地定位 300 户住宅。设计师负责现场分析、概念研究、土地规划、详细的分段计划以及通过市厅审批程序和社区作坊的详尽审查。根据市厅审批的设计方法，景观设计师负责设计、实施和所有公共区域、街道、样板房、入口墙壁和开放空间区域的景观的实地观察。

这个项目获得所有业主一致认可并且前所未有地受到了丹维尔规划委员会和市厅的一致认可。市长嘉奖了保留该地区几乎全部的 1000 棵橡树的独特成就。旧金山的郊区社区环保意识的在该案中反复使用，作为一个如何通过敏锐的土地规划来完成负责任的建筑的例子。

Verakin New Town

同景新城

Landscape Architect / SWA Group
Client / Chongqing Verakin Real Estate Co., Ltd.
Location / Chongqing, China
Area / 2,020,000 m²
Photographer / Tom Fox

The Verakin International New Town is a central component of Chongqing's New Tea Garden District. The project is located directly opposite the new District Government Center.

Three distinct zones characterize the site. The West Zone has largely been graded and appears horizontal and planar. The Center Zone is Stone Temple Mountain. In contrast with the West zone, The East Zone is sloped and terraced. A regional riparian open space corridor abuts the new town continuously along the south and east perimeters.

The master plan proposes higher density and active mixed use developments along the West, with tallest towers at the District Mixed Use Center along Century Boulevard as well as the riparian corridor to the South. From these perimeter locations, the architectural massing steps downward towards Stone Temple Mountain Park. Located at the center, the park is the central focus of the new town. On the East, lower densities are scaled to conform to the sloping terrain and massed to maximize offsite view opportunities. With all building massing, solar orientation and prevailing breeze considerations are integrated into the framework of the plan.

Community and neighborhood open spaces are provided at varying scales to offer citizens a wide range of active and passive activities. The open space amenities such as the Stone Temple Mountain Park, the regional Riparian Open Space Corridor, the District Green, the retail streetscape environments, and Neighborhood Open Spaces are strategically located to enhance accessibility and visibility.

同景国际城是重庆新茶园区的核心组成部分，与新区政府中心面对面。

该案由三个不同的区域组成。西区主要是分级的水平地形，中央区是石庙山，跟西区相对的东区是斜坡和梯田。沿着新城南边和东边的是一条河畔走廊。

总体规划中西区是高密度的综合建筑，包括沿着世纪大道以及向南的河畔走廊综合建筑区最高的塔群，建筑群朝向石庙山公园。位于中央区的公园是新城的焦点。在东区，缩减低密度以符合斜坡地形和集结最大区外景观的机会。所有的建筑群，阳光和自然风向的考虑都被纳入到计划框架中。

社区和邻里的开放空间，从不同范围为居民提供大范围的主导式和诱导式活动区。开放空间设施，如石庙山公园、河畔走廊、绿化区、商铺区、街景和邻里社区环境等的布置，增强了空间的可到达性与可视性。

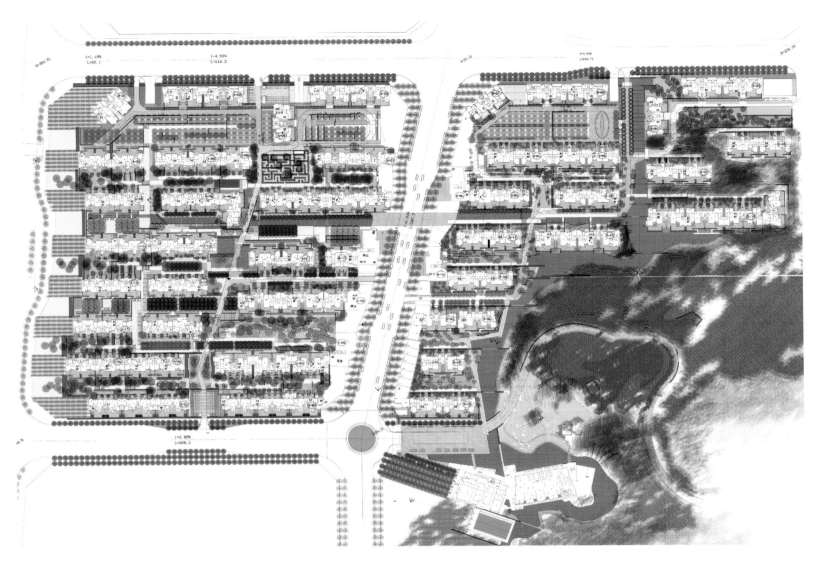

Vanke Longgang Mountain Living

万科龙岗山城小区

Landscape Architect / SWA Group
Client / Shenzhen Vanke Real Estate Co., Ltd.
Location / Shenzhen, China
Area / 200,000 m²
Photographer / Tom Fox

This residential community is a 20-hectare townhouse and condo development in NE Shenzhen. The Spanish-colonial town houses are surrounded by a lush landscape, complete with trails, pavilions and winding bike paths. Also a large majority of residents will live in high-rise condominiums. The income ratio between the town house and condominium residents is 10 to 1 so clearly there is great class division. For the sake of the community, a continuity of open spaces will benefit all and SWA's challenge was to provide a central green corridor and lake system accessible to all residents. The central open space connects both visually and physically (via bike paths) to a nature preserve immediately north of the property.

Although this project doesn't intend to get a LEED (Leadership in Energy and Environmental Design) certification, much of the design was inspired by LEED principles: maximize the open space to 50%, restore habitat in lake and woodland, reduce heat island effect by high density planting and pervious hardscape, encourage bicycle use, use regional material for construction, and reduce land development impacts from land cutting and filling.

该住宅小区占地20公顷，位于深圳东北部，项目包括联排别墅及公寓。西班牙殖民地建筑风格的联排别墅楼周边绿荫环绕，配备有小径、凉亭及蜿蜒的自行车道。相当大一部分居民则居住于高耸的公寓大楼中。居住于联排别墅和公寓大楼的居民收入比例为10:1，因此，它们在整体级别上有着较大的差别。为了整个社区的和谐，一个连贯的开阔空间无异将造福众人，而SWA（景观设计公司）的挑战在于如何规划出一个对所有住户开放的中央绿色走廊和湖泊。中央绿色空间无论在视觉上还是实际上（通过自行车道）都与这一地产北部的自然保护区直接相通。

虽然这个项目并不打算获得LEED（绿色能源与环境设计先锋奖）认证，但项目中许多设计的灵感都来自于LEED原则：将开阔空间面积扩大至50%，于湖泊林地中再造生物栖息地，通过密集种植和保留原始基础景观以减少热岛效应，鼓励使用自行车，在建筑中采用当地材料，在挖掘和填埋土地过程中减小土地开发带来的影响。

Nanshan Suzhou Golden Garden 1958

南山苏州金色花园 1958

Landscape Architect / Guangzhou Homy Landscape Co. Ltd.
Client / Suzhou Nanshan Real Estate Development Co., Ltd.
Location / Suzhou, China
Area / 113,000 m²

Natural and humanistic landscape elements were integrated into the landscape design, combining construction with the client's desire. The landscape design of the project was defined as Mediterranean style. Constructed with road system, the site was divided into three themes gardens — Italian, French, Spanish style. Each of three diverse themes has its own distinct characters at the same time is united under the main Mediterranean theme.

The design created a romantic, peaceful and harmonious community through the profound comprehension and innovation towards Mediterranean style, generating a leisure, friendly and natural Mediterranean style with humanistic environment which integrates the lines of urban constructions with the verdant freshness to make the owners feel the prosperous downtown, while enjoy a quiet life. The design realized the seamless integration between architecture and human being, as well as architecture and urban landscape.

我们把自然与人文景观元素融入环境设计中，结合建筑的设计与业主的意愿，将本项目的园林风格定位为地中海风格园林，设计以道路系统为骨架，把整个小区划分为意大利、法国、西班牙三大主题园林，这三个主题各具特色，同时又统一于地中海人文风情这一大主题之下。

本设计通过对地中海风格的深刻理解与创新突破，营造了一个浪漫、宁静、和谐的社区氛围，再现一种休闲写意、亲切自然的地中海风情人文环境，将都市建筑的线条和绿色生态的清新融为一体，既让业主感受城市的繁华，又能拥有宁静的生活，实现建筑与人、建筑与城市景观的完美融合。

Chevron Garden

雪佛龙花园

Landscape Architect / Eckersley Garden Architecture
Location / Melbourne
Photographer / Sarah Appleford

The concept of Chevron garden was to enhance the experience of everyone living within the environment. It is intended that spaces are friendly, functional and relaxed, yet they are designed with attention to detail that would suit any discerning style advocate.

Special elements are incorporated to add to the sensory pleasure enjoyed by the people using the spaces. So whether just passing through, or stopping to relax, something of the gardens should leave a positive and inspiring impression.

雪佛龙花园的设计理念是提高环境中每个人的生活体验。它所追求的是，空间是：友好的、功能性的、让人放松的，且这些空间的设计注重细节，能适应任何所倡导的可识别风格。

本案设计采用了特别元素，增加空间使用者的感官享受。所以无论是路过，或是停下来放松自己，花园里的一些东西应该能留下积极和鼓舞人心的印象。

Villas del Mar at Palmilla

德尔玛别墅

Landscape Architect / SWA Group
Client / Careterra Transpeninsular
Location / San Jose Del Cabo, Mexico
Photographer / Tom Fox

Villas del Mar is an exclusive residential resort community with hotel, golf course and other amenities, located in Los Cabos, Mexico. SWA provided planning for three phases of the development, overall landscape design, and the design for the individual yards for each of the units. In total, the project includes about a hundred attached units that overlook the Bay of Los Cabos and the Sea of Cortez. Two of the phases consist of million dollars zero lot line homes lining the beach. Each home is individually designed with beach access and a private, infinity edge pool. The third phase includes steep slopes and rocky conditions as well as spectacular views of the ocean. The design integrated the units into the mountainside without destroying the mountain's natural features. Native desert plant materials were used and dramatic fire torcheres were incorporated to light the steep roads at night.

位于墨西哥洛斯卡沃斯的德尔玛别墅是拥有酒店、高尔夫球场所和其他设施的高级住宅度假村。SWA为整体景观设计规划了三个阶段，并为每单元每一户都提供了设计。总的来说，该案包括100套连接的单元，俯瞰着洛斯卡沃斯海湾和科尔特斯海。两期由价值百万美元的小独栋别墅组成，每一栋都是单独设计，可直达海滩，配备一个私人的无尽头的泳池。第三期包括陡峭的斜坡和岩石以及大海的壮观景色。该设计将住宅融入到山坡上，而没有破坏山的自然特征。采用了原生沙漠植物和引人注目的高火塔，在夜间照亮陡峭的道路。

Sierra Bonita Mixed Use Development

Sierra Bonita 综合大楼

Landscape Architect / AHBE Landscape Architects
Client / West Hollywood Community Housing Corporation and Tighe Architecture
Location / West Hollywood, California, USA
Area / 4,645.15 m²
Photographer / AHBE Landscape Architects

Completed in 2010, this new 42-unit West Hollywood establishment provides low-income affordable housing for those with special needs. Designed by Tighe Architects, this award-winning and highly publicized project is a pilot for the City of West Hollywood's Green Building Ordinance, one of the first ordinances of its kind in the country. The project includes a solar electric panel system integrated into the facade and roof of the building. The wrap around central courtyard is planted with bamboo to provide shade and higher air quality for the residents. Drought resistant plantings use drip irrigation from rainwater while concrete enclosures act as planters to filter the rainwater before it leaves the site. AHBE's landscape design helps to both increase the project's long-term sustainability as well as connect the building, which has a striking urban presence, to its surrounding environment.

2010年完工的，这栋新的42单元的西好莱坞建筑为那些低收入有特殊需求的人提供了经济适用房。由Tighe建筑事务所设计的，这个屡获殊荣的备受公众关注的项目是全美首创的条例之一的西好莱坞绿色建筑条例的先锋项目。该项目包括一个融入到外墙和楼顶中的太阳能面板系统。围绕中央庭院的是竹林，为居民提供遮荫和较高质量的空气。耐旱植物使用雨水滴灌，混凝土外壳的花坛，过滤流过该地的雨水。AHBE（景观设计公司）的景观设计协助增加项目的长期可持续性以及将具有鲜明都市特色的建筑与其周围环境联系在一起。

South Park Streetscape and Mixed Use Development

南园街景和综合高层大楼

Landscape Architect / AHBE Landscape Architects
Client / The South Group Partnership
Location / Downtown Los Angeles, California, USA
Area / 18,377.24 m²
Photographer / Heliphoto / Jack Coyier

Located in a four-block area in Downtown Los Angeles' South Park neighborhood, the project consists of three mixed-use LEED Gold certified buildings (called Evo, Luma, and Elleven), the streetscape, and a common open space area associated with each building.

AHBE's streetscape is unique to Los Angeles, designed to create a new sense of community by encouraging pedestrian activity around the area, and leaving space for outdoor eating for sidewalk cafes as well as shaded benches to encourage lingering.

The streetscape design also features planters specifically designed for storm water infiltration. During a storm, rainwater flows from the street into the gutters and is then diverted into the infiltration planters, by way of cuts in the curb, and percolates back into the ground. This system minimizes the amount of storm water run-off into the City's system, cleans the water through a natural process and recharges the region's aquifers. This project is a key precedent for the City of Los Angeles in developing its own Green Streets program, as well as the model for implementing sustainable design with the infrastructure.

South Park's design further reinforced the sense of community by creating several open space areas for the residents of the buildings. Each high rise building includes an open courtyard space. Between Evo and Luma is a courtyard area that is open to the public during daylight hours. A rooftop outdoor living/seating/eating/cooking area is accessible to the residents of Luma, Elleven, and Evo and features a pool deck and sky patio.

This project exemplifies the possibilities of green building and landscape design throughout Southern California.

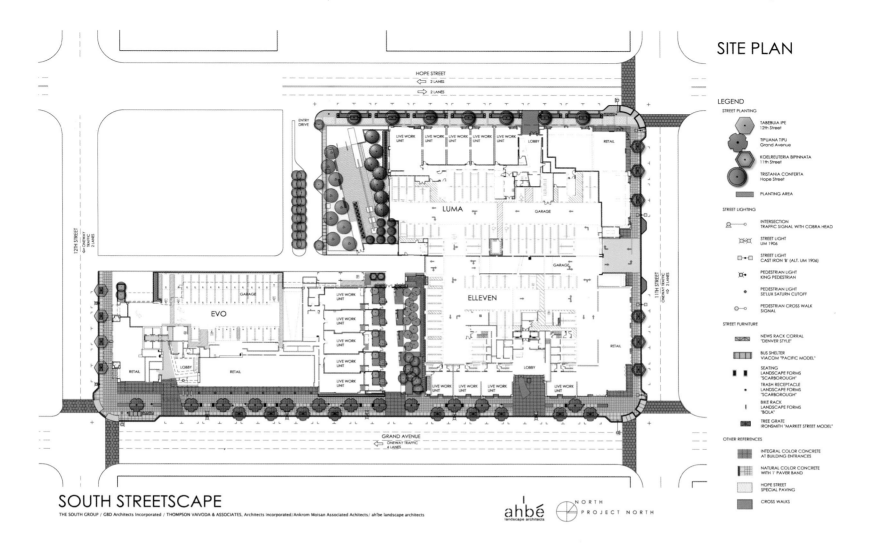

SOUTH STREETSCAPE

THE SOUTH GROUP / GBD Architects Incorporated / THOMPSON VAIVODA & ASSOCIATES, Architects Incorporated/Ankrom Moisan Associated Architects/ ah'bé landscape architects

该案位于洛杉矶市中心南园附近的一片四街区域，其由三栋获得LEED金级资格认证的综合高层大楼（名称分别是Evo、Luma和Elleven）、街景以及与建筑有关的公共空间构成。

AHBE设计的街景对洛杉矶而言是独一无二的，该设计通过鼓励行人在周围活动，留出户外就餐和露台咖啡馆的空间以及鼓励停留的荫凉长椅，创建一个社区新意。

街景设计还包含专为雨水渗透而设计的花盆。在暴雨中，雨水从街道流入排水沟，然后被转移到渗透花盆中，经过边石削弱，渗透到地面。该系统最大限度地减少雨水排放到城市系统，通过自然过程净化雨水，补充该区域的地下水层。这个项目是洛杉矶市在开发绿色街道规划中的一个重要范例，也是基础设施实施可持续性设计的模型。

南园的设计，通过为建筑居民创造几个开放的空间进一步增强了社区意识。每栋高层建筑都有一个开放的庭院空间。在Evo和Luma之间是白天向公众开放的庭院区。Luma、Elleven和Evo的居民可以到达屋顶的室外生活区、座椅区、餐饮区、烹饪区以及游泳池甲板和空中阳台。

该项目充分体现了南加利福尼亚州绿色建筑和景观设计的可能性。

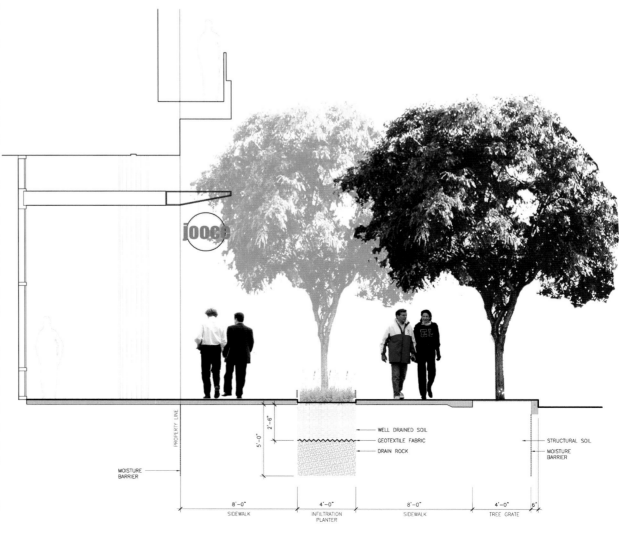

Lander Taizhou LaiYinDongJun

莱茵达泰州莱茵东郡

Landscape Architect / Guangzhou Homy Landscape Co. Ltd.
Client / Taizhou Lander Real Estate Co., Ltd.
Location / Taizhou, China
Area / 45,000 m²

The architectural design for the project is typical modern style with the concise and grand Singaporean landscape as the design theme. While the landscape design coordinated with the overall planning and architecture, the concept of "modern, harmonious, ecological, recreational, practical" was integrated into the landscapes. With the intention to bring a high quality living space manage to create a low density, ecological, and healthy living environment which fully represents leisure, relaxed, comfortable life pursuit. At the same time it embodied the environment with functions to improve the artistic quality of environment, allowing people not only feel the physical and mental relaxation and pleasure, but also receive spiritual edification.

该项目的建筑风格为典型的现代风格，以新加坡园林作为本设计的主题，简洁而大方。园林景观设计在与整体规划及建筑风格协调的同时，将"现代、和谐、休闲、生态、实用"的思想融合到景观中，从创造人居高品质环境的目的出发，力求创造一处低密度、生态、健康的居住环境，充分体现休闲、自由、舒适的生活追求。并在赋予环境功能性的同时，升华环境的艺术品质，使人们不仅能感受身心的放松与愉悦，更能受到精神的陶冶。

Kuany – Huizhou Holland City

光耀——惠州荷兰小城

Landscape Architect / Guangzhou Homy Landscape Co. Ltd.
Client / KUANY Group
Location / Huizhou, China
Area / 60,000 m²

The landscape design subjects to the theme of "Dutch Flavor". The Netherlands is a country which clusters elegant artistic atmosphere and unique living culture. Based on the lavish artistic aura, the entire site sticks to Dutch artistic elements as the principal part, with living culture as the complementary one, Dutch style landscapes are reflected in the pavement, sculptures, plants and etc. via practical approaches, so as to maximize the views from visual, auditory, tactile, olfactory to experience the full range of the most natural and vital views in this Holland Town. Plantings, lakes, streams, fountains and other landscape elements are used to combine with the minimal European style, along with Dutch elements such as windmills, tulips to create a distinctive unique Holland town making people fall in love with the town from the bottom of their hearts. The project has won "The Best Landscape Real Estate Development Award" in China.

本项目景观设计以"荷兰风情"为主题，荷兰是集聚着高雅的艺术气息与特有的生活文化为一体的国家。借着浓浓的艺术氛围，整个居住区以荷兰艺术元素为主导，以生活文化为辅线，通过可实施的方法将荷兰特色景观体现在铺装、雕塑、植物等元素中，从而最大限度地让人们从视觉、听觉、触觉、嗅觉全方位感受到荷兰小镇的最自然、最富有生命力的艺术景观，同时设计巧妙运用植物、湖泊、溪流、喷泉等景观元素，结合简约欧式风格，并加入风车、郁金香等荷兰主题元素，创造出别具特色的荷兰小城。让人们由衷地爱上荷兰小城。该项目获得"中国房地产最佳园林景观楼盘奖"。

Campbell, Salice & Conley Residence Halls

坎贝尔、萨利斯 & 康利学生宿舍楼

Landscape Architect / Sasaki Associate
Client / Fordham University
Location / Bronx, NY, USA
Area / 15,793.52 m²
Photographer / Robert Benson

The residence halls at Fordham are the first project implemented from Sasaki's framework plan for the campus. The buildings are situated at main pedestrian entry to campus, near the intersection of a major city boulevard and a commuter rail line. The structures establish a celebratory gateway to the campus from the Bronx. A historic pedestrian path from the gateway is accentuated with a series of open spaces. Gradually, this public zone gives way to varied paving patterns, more trees, and a more intimate green space that includes a quiet courtyard. Adjacent to several existing residence halls, the project effectively consolidates a residential neighborhood on the west side of campus, and provides an opportunity for the area to become a mixed-use, student life hub.

The new buildings are set on raised terraces, and establish a sense of place by framing one of the most important green spaces on campus. Each of the two buildings is articulated as two towers with a shared lobby. This strategy supports smaller student neighborhoods at each floor, reinforcing the university's focus on community. At the heart of residential floor, double-height lounges offer opportunities for socialization and group learning. At the first floor, the buildings house a café, a multipurpose room, and two integrated learning centers — key components of the halls' living & learning programs. Even the laundry room — an often overlooked space — is designed as a social place,

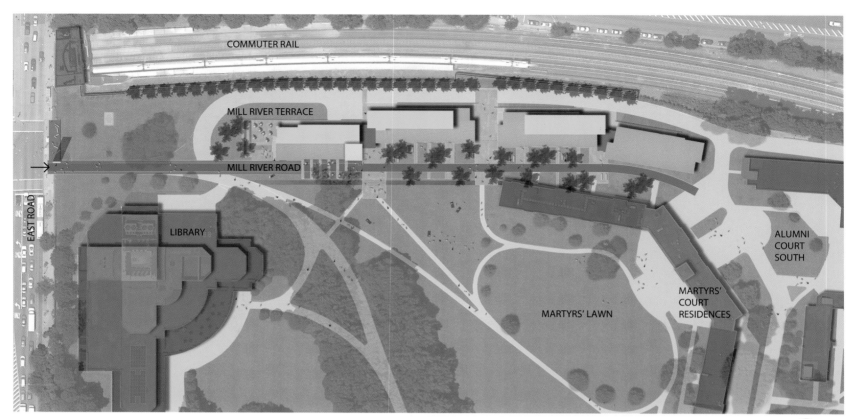

SITE PLAN

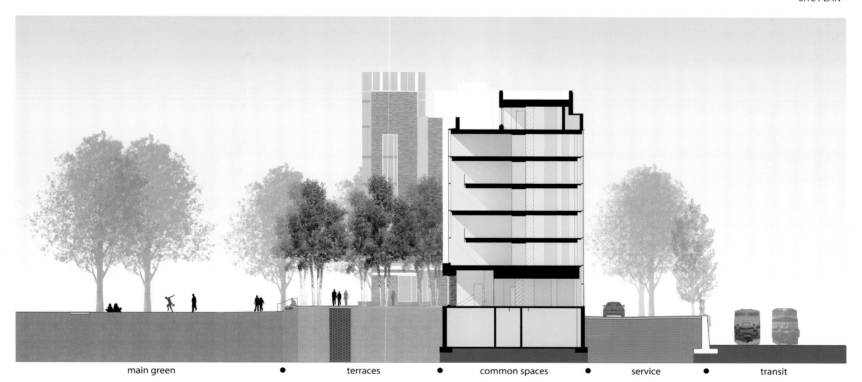

SECTION

located adjacent to casual study areas and with direct visual connection to the exterior via a glazed arcade.

As part of the university's ongoing commitment to sustainability, the new residence halls take advantage of natural light and ventilation. Highly porous paving is incorporated into the site design and integrated with a major new storm water detention system to reduce water infiltration issues. The project has achieved LEED® Gold certification.

在福特汉姆的住宅区是Sasaki（景观设计公司）在校园框架实施的第一个项目。建筑群位于校园主人行道的入口处，靠近一城市林荫大道和城市铁路的交叉处。这些结构使通道从布朗克斯通到了校园里，由此形成的空间突显了一条具有历史意义的人行道。这个公共区，为不同的铺装图案，更多的树木，和包括一个静谧的庭院在内的更亲密的绿色空间提供了空间。毗邻几栋现有的宿舍，该项目有效地巩固了校园西侧的住宅周边区域，并为该区域提供了一个成为综合大楼——学生生活中心的机会。

新建筑物设置在阶梯上，通过框住校园里最重要的绿色空间来制定空间感。每两栋建筑物相连，成为共用一个大厅的两座塔。这一战略在每一层都支撑着较小的学生社区，加强了社区的大学中心。在住宅楼的心脏区，双层高度的休息室提供社交和小组学习的机会。一楼有咖啡馆、多功能室和两间综合学习中

心，学习中心是大厅生活和学习的关键组成部件。甚至洗衣间，一个经常被忽视的空间，也被设计为一个社交的地方，毗邻休闲学习区，通过玻璃拱廊在视觉上直接连接到户外空间。

作为佛特汉姆大学可持续发展的目标的一部分，新宿舍建筑利用自然光和通风，将多孔铺路纳入设计，融合了主要的雨水收集器，减少雨水渗透问题。本项目已获得 LEED® 金级资格认证。

California State Polytechnic University, Pomona

加州州立理工大学住宅区，波莫纳

Landscape Architect / Sasaki Associates
Client / California State Polytechnic University
Location / Pomona, CA, USA
Area / 35,844.04 m²
Photographer / Bruce Damonte, R. Greg Hursley

The residential suites compose a residential district that centers student life near the academic core. The 16-acre community features more than 1,000 beds, commons buildings offering food service, a convenience store, laundry facilities, resident parking, and various recreational amenities. The design focus was to enhance community by creating a neighborhood-like setting.

Each suite has shared interior spaces comprised of a living area, kitchen center, storage closet, and two compartmentalized bathroom facilities. Shared balconies and common study spaces are clustered at prominent building corners - available for groups, individuals, and seminars. Aligned with vertical circulation, these stacked commons spaces act as glowing beacons at night, providing residents with a sense of identity and orientation.

Primary pedestrian paths for nonresidents are routed around the project's perimeter. These edges are defined by the buildings, which vary from three to four stories in height. Internal community spaces shaped by the buildings take the form of a series of linked courtyards and open spaces. The courtyard spaces are designed for formal and informal housing functions, campus programs, and student life recreation. Sasaki maintained a significant grove of historic sycamore trees and incorporated them into the project, creating a gateway into the residential community from the new quad. The team also developed a pedestrian promenade that links the academic core to the new quad, athletics, the housing precinct, and a commuter parking area.

An integrated approach to sustainability is reflected in the sensitive sitting and orientation of the buildings, the shading of south- and west-facing glazing, an energy-efficient central plant, and careful selection of interior materials and finishes. Phase II of the community is LEED® Silver Certified.

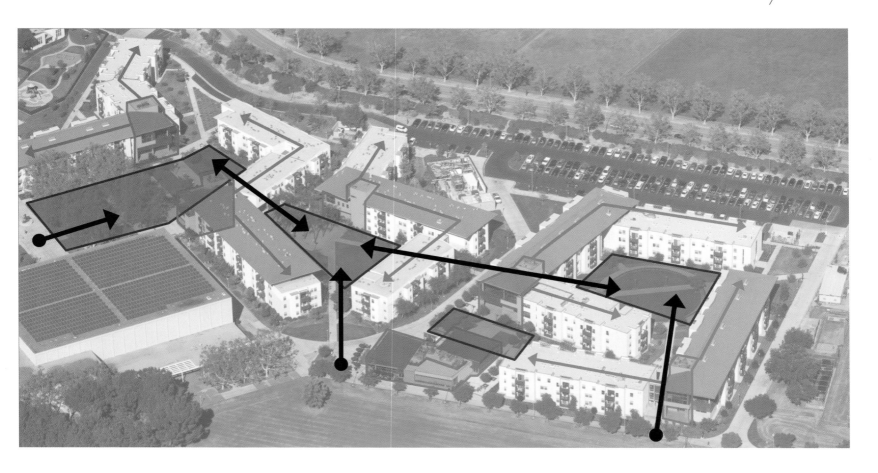

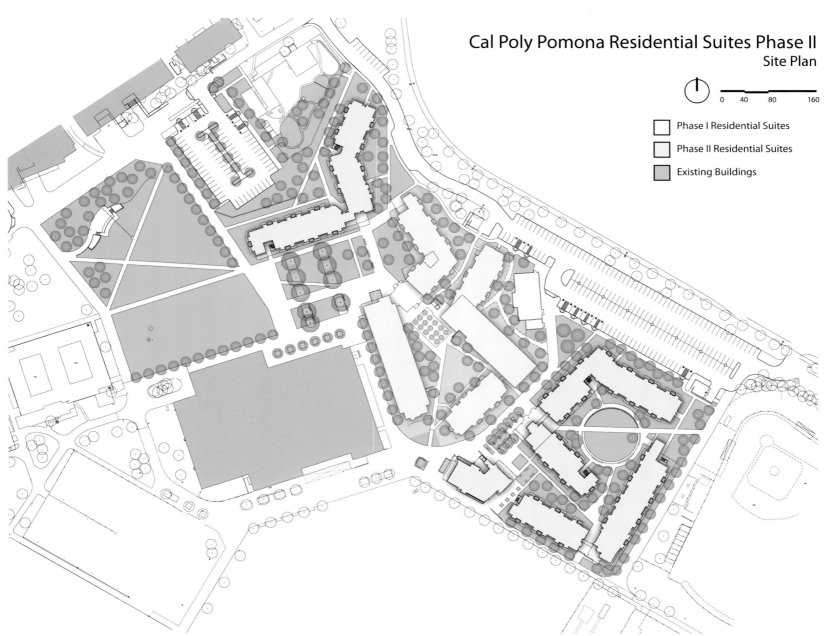

Cal Poly Pomona Residential Suites Phase II
Site Plan

- Phase I Residential Suites
- Phase II Residential Suites
- Existing Buildings

　　该套房区组成的住宅区将学生生活集中在学校核心区附近。16 英亩的社区包括了 1000 多张床位和提供餐饮、便利店、洗衣、停车场以及各种娱乐设施的公共建筑。设计侧重于通过创建一个类似社区的区域以提升校园社区质量。

　　每个套间有公用的内部空间，由生活区、厨房、储藏柜和两个分隔的浴室组成。公共阳台和学习空间集中在建筑物的角落，供团体、个人和研讨会使用。均衡垂直循环体系，让这些堆积的公共空间在夜间作为发光的灯塔，方便居民辨认和定位。

　　非居民的主要人行道围绕着项目周边。这些边缘被建筑物所定义，高度上

有三到四层的变化。建筑共享的社区内部空间采取的形式是相连的庭院和开放空间。庭院空间是为形式和非形式的住房功能、校园计划和学生生活娱乐而专门设计的。Sasaki保留了具有历史意义的梧桐树林并将其纳入项目中,形成了一条通道,连接学校中心和新住宅社区。设计小组还开发了一条人行长廊,连接学院区和新的方形院子、体育中心、住宅区和停车场。

灵敏的坐落位置和建筑物的方位反映了可持续发展的策略:南向和西向玻璃幕墙的阴影,一个中央发电机和精挑细选的室内装饰材料。第二阶段是申请LEED® 银级资格认证。

UCLA Northwest Campus Redevelopment

加州大学洛杉矶分校西北校区重建项目

Landscape Architect / SWA Group - Richard Law, FASLA, ULI, Principal
Client / UCLA and Pfeiffer Partners Architects, Inc.
Location / Los Angeles, California
Photographer / Tom Fox, Principal, SWA – SWA Field Group – Denise Retallack, SWA

As a part of UCLA Northwest Campus Redevelopment master Plan, SWA was engaged for pre-design and full service landscape architectural services for DeNeve Residential Towers and Sproul Complex, including three new residential towers and a new mixed-use facility consisting of a multi-level residential structure, a large ballroom with lobby, meeting rooms and support space, a 750-seat dining commons, the Office of Residential life and maintenance offices with parking. Also included are on-structure landscaped terraces, plazas for exercising and gathering, open spaces, exterior stairs, site lighting and pedestrian pathways.

SWA also provided full landscape architectural services for the renovation and redevelopment of Hedrick and Rieber Precincts of the Northwest Campus. This project included the redesign and construction of undergraduate housing with an additional 2,000 beds, a 248-car parking structure and associated academic, recreational and student programming facilities designed to integrate with the residential living experience, and provide an environment-rich combination of spaces and activities that foster the vitality in on-campus residential life.

作为 UCLA（加州大学洛杉矶分校）西北校区重建总规划的一部分，SWA 参与了迪维住宅塔楼和斯普劳尔综合楼的预先设计和全面景观设计，包括三栋新的住宅塔楼和一栋新的综合大楼。综合大楼由多层住宅、配备大厅的大型宴会厅、会议室和对应空间、一个 750 座的公用餐厅、居住生活办公室和维修办公室及停车场组成，还包括景观阳台、运动和聚会的广场、开放空间、户外楼梯、照明和人行道。

SWA 还为西北校区的赫德里克区和里伯区的重建与重新改造提供整套景观建筑设计。该项目包括重新设计和建造一个本科生宿舍，有 2000 个床位、248 个停车位的建筑和相关的学术、娱乐和为学生设计的设施，设计成与居民生活体验相结合，并提供空间和活动环境丰富的结合，在校园生活中培养活力。

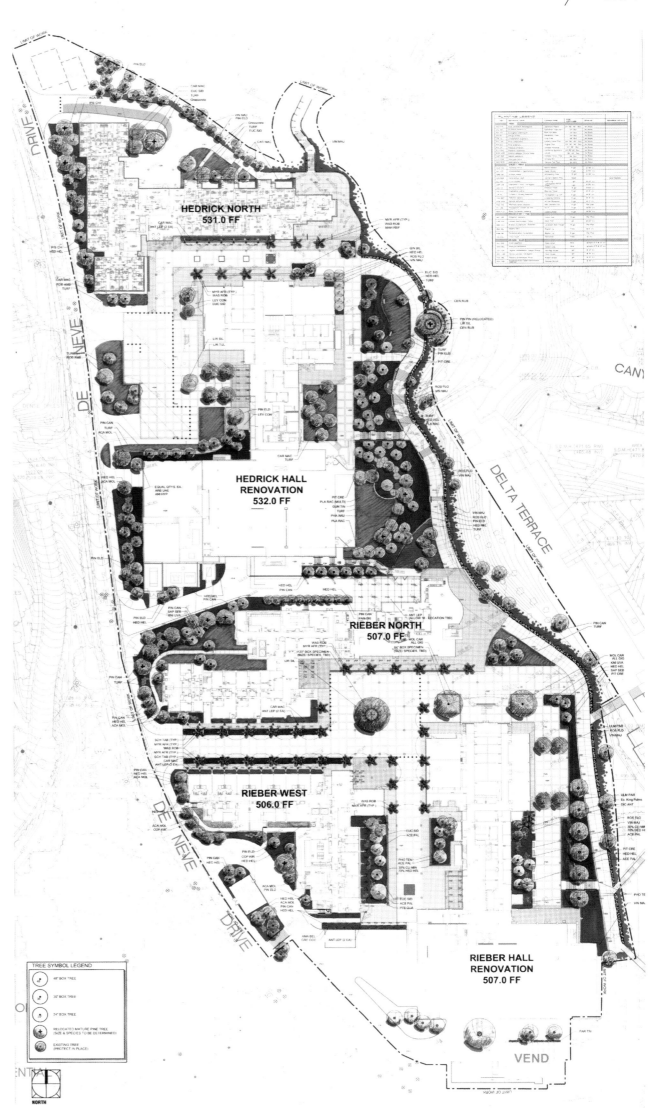

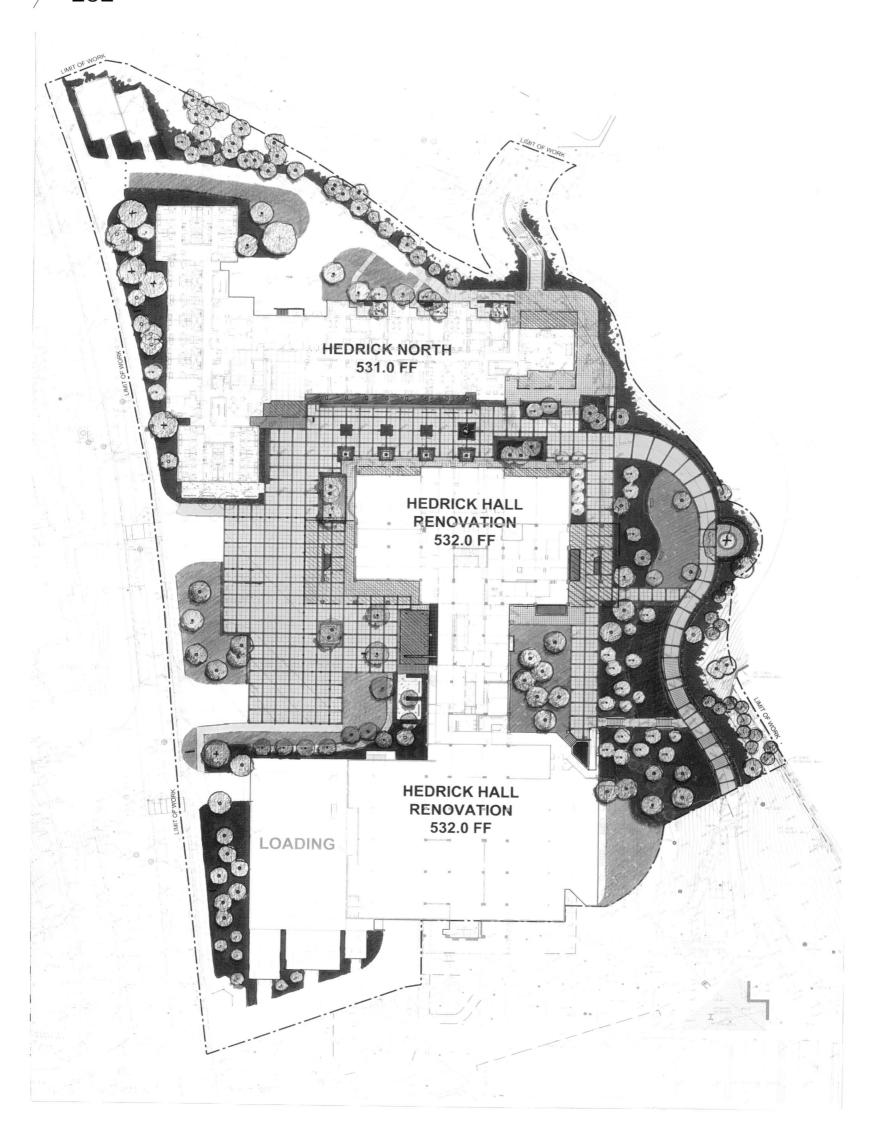

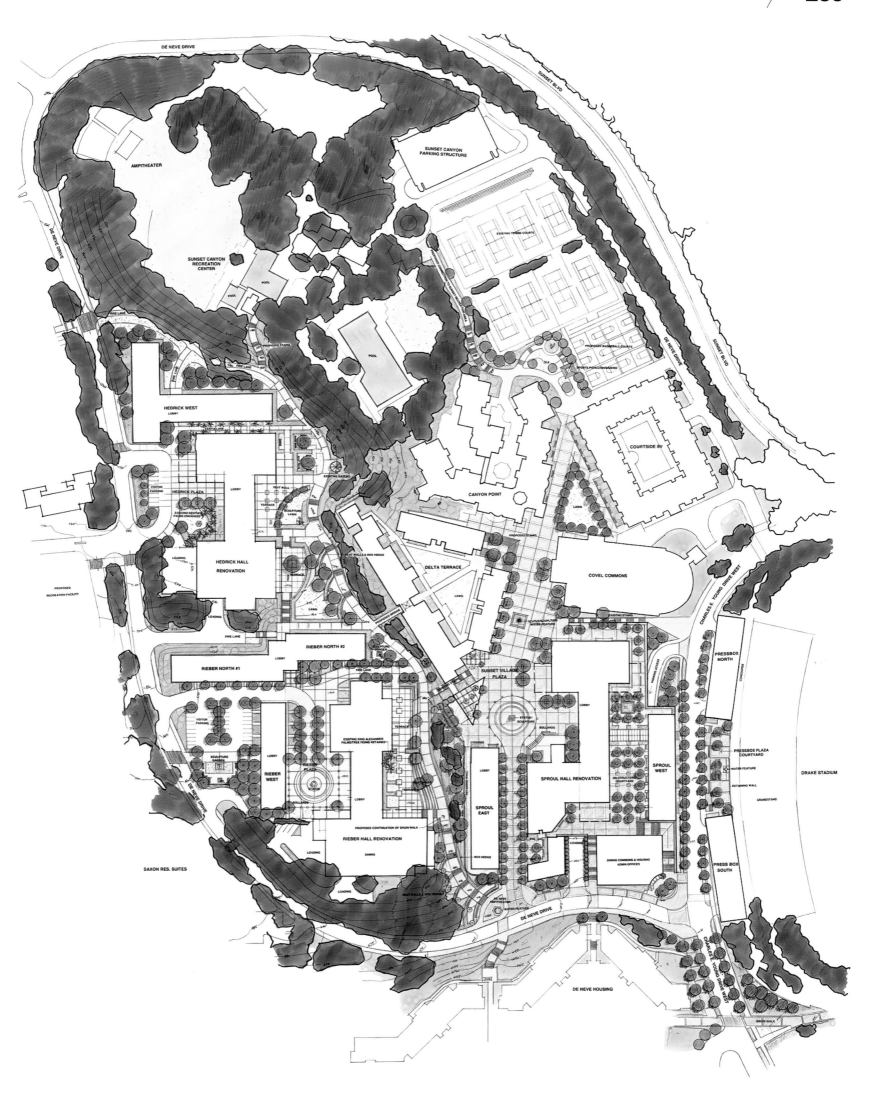

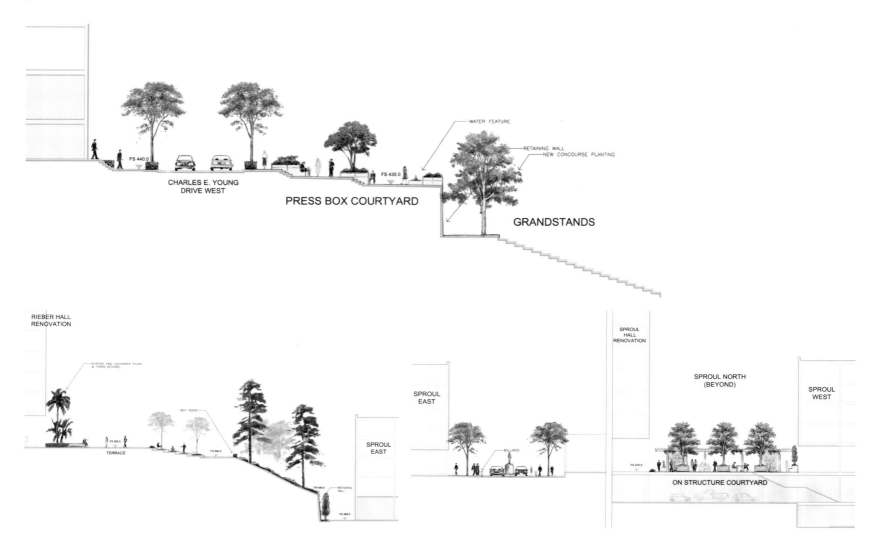

UC Davis West Village Phase I

戴维斯大学西村第一期

Landscape Architect / SWA Group – John L. Wong, FASLA, FAAR, Managing Principal, Chairman
Client / West Village Community Partnership, LLC
Location / Davis, California
Area / 809,371.28 m²
Photographer / Jonnu Singleton, SWA Group, Ken Cantrell, Viewpoint Aerial

The West Village Master Plan was a response to the substantive growth in the number of students, faculty and staff on the Davis campus, and rapidly escalating housing costs, which together have forced campus affiliates to seek housing outside of Davis. The West Village residences and apartments provide new choices for those who desire to live, work, and recreate in a sustainable residential neighborhood that is seamlessly integrated with UC Davis' core activities. The implementation plan is based on three core principles: housing affordability, environmental responsiveness, and quality at place.

The extensive design elements that make this the largest Zero Net Energy Community in the US:

- Compact walkable Neighborhoods
- Stormwater Detention and Retention Ponds
- Extensive Bike Network
- Environmentally Responsive Architecture
- Permeable paving
- Photovoltaic Panels
- Solar Thermal Collectors
- Passive Cooling of Building and Pavements
- Extensive Tree Shade

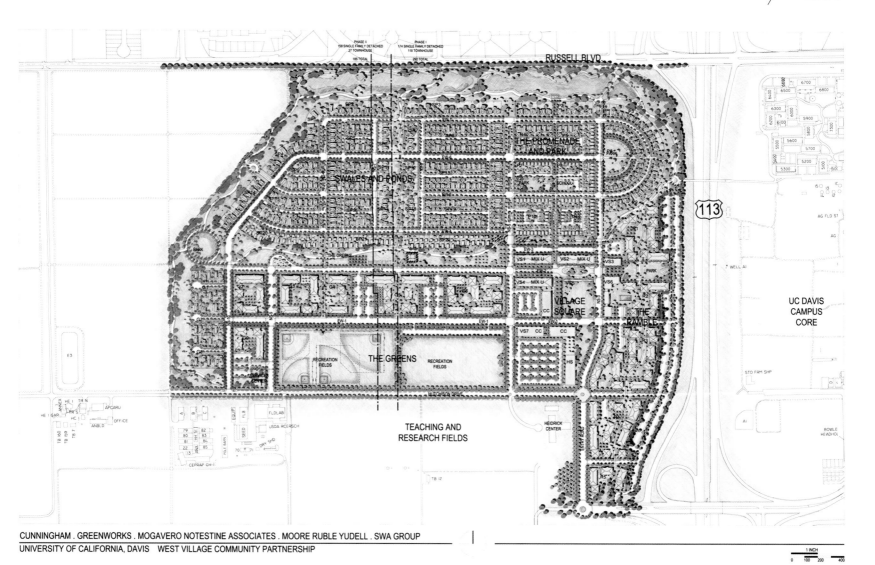

CUNNINGHAM . GREENWORKS . MOGAVERO NOTESTINE ASSOCIATES . MOORE RUBLE YUDELL . SWA GROUP
UNIVERSITY OF CALIFORNIA, DAVIS   WEST VILLAGE COMMUNITY PARTNERSHIP

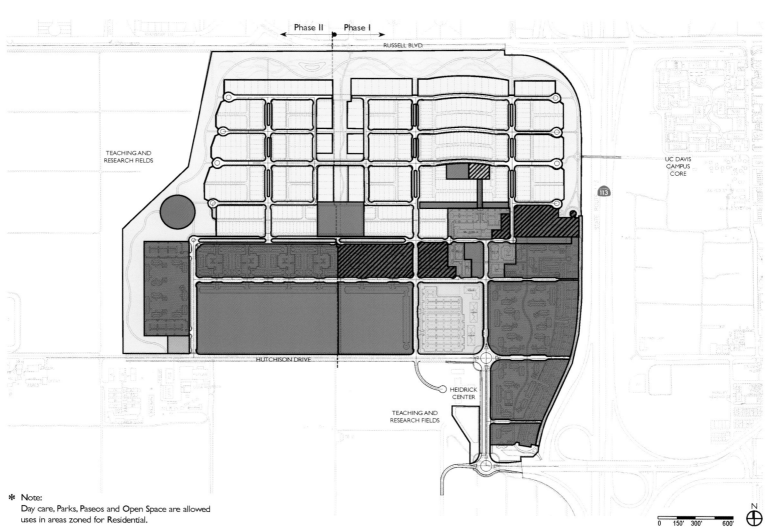

* Note:
Day care, Parks, Paseos and Open Space are allowed uses in areas zoned for Residential.

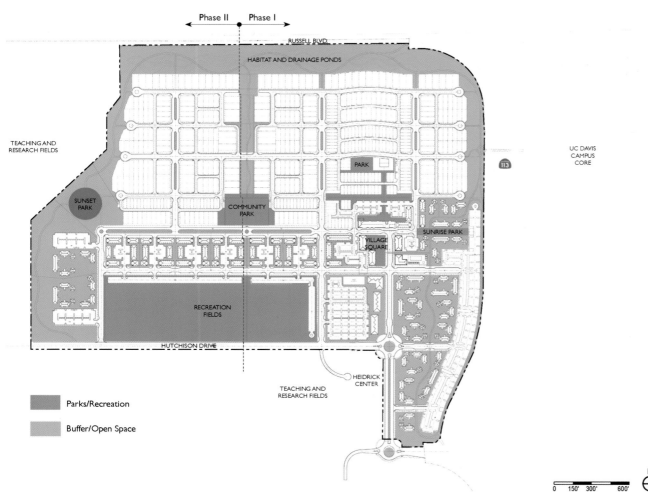

西村总体规划是针对戴维斯校区的学生、教师和职工数量的实质性增长以及迅速攀升的房价，这迫使校方寻求戴维斯校区以外的住宅。西村住宅和公寓为那些渴望在可持续的住宅区生活、工作和消遣的人们提供了新的选择，该住宅区完美融入戴维斯大学的核心活动中，实施方案基于三个核心原则上：住房负担能力、环境反应和空间质量。

大量的设计元素让这里成为美国最大的零净能源社区：

- 紧凑的步行社区
- 雨水调蓄池
- 大规模的自行车网道
- 环保建筑
- 透水铺面
- 光伏板
- 太阳能集热器
- 建筑和路面的诱导式冷却系统
- 大量的树荫覆盖

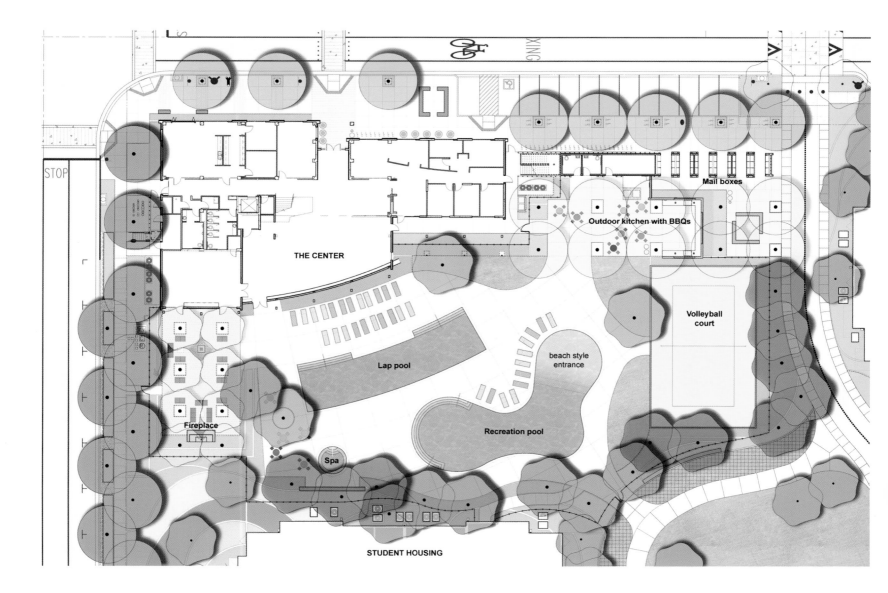

Lantern Bay Master Plan and Coastal Design

灯笼湾总规划和景观设计

Landscape Architect / SWA Group
Client / Smythe Brothers, Inc. and Pacific Mutual Life Insurance Company
Location / California, USA
Area / 310,000 m²
Photographer / Tom Fox

SWA provided master planning and landscape architecture for this 76-acre coastal park and multiuse project that rehabilitates the site from a scarred, abandoned borrow pit to a coastal blufftop open space of great value to Southern California residents and visitors.

The Blufftop Parks are two interrelated public parks that form the central open-space element of SWA's master plan, situated on the coastal bluffs overlooking Dana Point. SWA integrated the parks with commercial, hotel, and residential uses and organized public workshops to identify community concerns while developing the land use program. To meet requirements for disabled access SWA's grading plan created a series of semicircular switchbacks that traverse up the 120-foot-high bluff. The resulting land sculpture rewards walkers with dramatic views of the Pacific Ocean from resting places at the end of each terrace.

SWA为这个76英亩的海滨公园并具其他功能的项目提供总体规划和景观设计，将这个坐落于海滨悬崖边上、满目疮痍且被废弃的取土坑复原为一个宝贵的、可供南加州居民和游客游玩的开放空间。

该崖顶公园坐落在海安的峭壁上，俯瞰着丹纳岬，由两个相互关联的公园组成，构成了SWA的总体规划中的中央户外元素。SWA在开发该土地的同时，使得公园与选址的商业、酒店、住宅各种用途合为一体，并可在此举办公共研讨会来探讨社区问题。为了方便残疾人到此一游，SWA的分级计划打造了一系列半弧形的盘山路，一直延伸至36.6米高的悬崖。由此产生的陆地刻蚀为步行者提供了壮观的视野，在每个阶地末端的休息处都能看到太平洋。

ILLUSTRATIVE PLAN

LANTERN BAY
Dana Point, California

THE SWA GROUP

Xinsheng Oriental Blessing

杭州欣盛东方福邸景观设计

Landscape Architect / L&A Design Group
Client / Xinsheng Real Estate Development Co., Ltd.
Location / Hangzhou, China
Area / 113,776 m²

The project showcases an idyllic scene of the modern natural landscape garden in Hangzhou. However, the real intention is to highlight the so-called "predictability yet implicitness", that is, through the images of materialized inner emotions, philosophical experience and links to realize the unique design. The prospect of a pine tree, a couple of plexus and delicate flowers in the garden outlines the artistic conception which makes people have the feeling of "the heart knows the wonder scenes while mouth can't describe it".

Design Concept:

The temperament and subtlety of Hanzhou gardens lie in the broad and profound cultural roots. The project referred to the native landscape design concepts and deduced with contemporary gimmick, creating a modern new landscape with fresh air, warm sunshine, rhythmic stream and forested virescence.

Garden District buildings face the water and rest on the hill. The unique design of annular households reflects the sense of respect and superiority. The encircling overhead layer links the three high-level units, as the walkway to protect from rain and wind. Indoor lobby where one can enjoy the views and outdoor pavilion face each other in distance. The exterior landscape and interior overhead layer interact with each other. The waving mountains strengthen the depth of the garden. The rich plantings' combinations indicate the natural elegant living taste, as well as the beautiful landscape of lakes and mountains, bringing visitors carefree and relaxing experience.

该项目充分展示了杭州现代自然山水园林所富有的诗情画意。但造园者的真正目的是强调所谓"意向之象,言外之意",就是通过物化的内心情感、哲理体验和链接独特的形象联想来实现设计。园林中的一株青松、几丛翠绿、少许娇花所勾勒出的意境便使人有"所致得奇妙,心知口难言""言有尽而意无穷"之感。

设计理念

杭州园林具有如此的气质和精妙在于其博大精深的文化根源。该案参照杭州本土化的造园理念,以现代手法重新演绎,打造全新的杭州现代山水,兼有清新的空气、和煦的阳光、灵动的溪水和森林般的绿化。

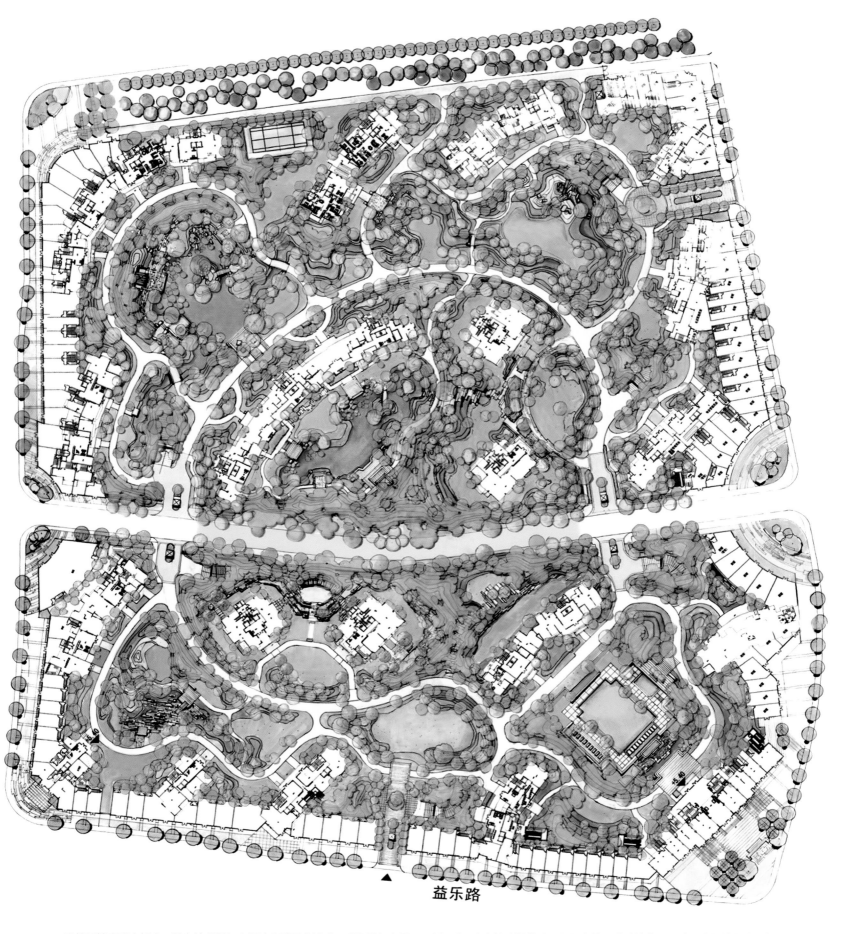

益乐路

园林区楼房面山望水,独有的环形入户设计体现了尊崇感,环抱型架空层将三个高层单元连接,如同风雨长廊。室内观景大堂与户外景亭遥相对望,通过幽幽小径使户外园景与室内架空层形成互动。起伏环绕的山峦加强了园林的幽深感,丰富的种植搭配,尽显自然、雅致的生活品味,优美的湖光山色,让置身其中的访客感受悠然自得的情趣。

Southport Central

南港中心

Landscape Architect / PLACE Design Group Pty Ltd.
Client / Raptis Group
Location / Southport, Gold Coast, Queensland Australia
Photographer / PLACE Design Group Pty Ltd.

Southport Central is a, 3 tower, mixed-use development seamlessly uniting luxury residential apartments, premium office suites and a ground floor retail precinct.

Located within Southport, the Gold Coast's CBD, Southport Central is ideally positioned to take advantage of surrounding facilities, including Australia Fair and the Southport Library, etc., the planned translink transportation route and the upgrade of Smith Street, enabling easy access to Brisbane.

LIVE — Comprising 1+1, 2+1 and 3 bedroom luxury apartments with panoramic Broadwater and hinterland views and resort style facilities, Southport Central offers residents a luxurious lifestyle.

WORK — Located within the Gold Coast's CBD, Southport Central is set to become a leading commercial centre, with networking opportunities within modern surroundings. Office suites can be configured to suit any business needs, with suite sizes from 50m^2 up to independent buildings.

PLAY — The ground floor plays the host to a vibrant retail precinct, with fashion boutiques, cosmopolitan dining options, beauty salons, services and, soon, a fresh food emporium.

南港中心是一栋三塔式综合高层建筑，集豪华住宅公寓、高档写字楼和地面商铺空间于一体。

位于黄金海岸中心商务区的南港，南港中心的理想位置可充分利用周边设施，包括澳大利亚会展中心和南港图书馆等，便利的交通路线和整修后的史密斯大街都能够方便地通往布里斯班。

生活——包括1+1，2+1和3卧室的豪华公寓，可看到布罗德沃特腹地的全景，配备度假村风格的设施，南港中心为居民提供了一种奢华的生活方式。

工作——位于黄金海岸中心商务区内，南港中心将成为一个主要的商务中心，加上现代的周边与交通网。写字楼套间能适应任何商务的需求，套间面积从50平方米到独立建筑都有。

玩乐——地面层是一个充满活力的商铺区，包含时装精品店、国际化的餐饮、美容院、服务区，还有一个即将落成的生鲜食品商场。

Cutters Landing

卡特斯码头

Landscape Architect / PLACE Design Group Pty Ltd.
Client / MIRVAC
Location / Brisbane, QLD, Australia
Photographer / PLACE Design Group

PLACE Design Group was commissioned to carry out the landscape consultancy work for Cutters Landing. It is an excellent example of contemporary minimalist urban design, it is one of the city's newest riverfront development.

The landscape color language and geometry complements HPA's architecture and at times contrasts to announce features to the wider area. A fig arbor was established as a pedestrian thoroughfare link to the extended river boardwalk to the Powerhouse theatre and restaurants.

The riverfront swimming pools are designed with infinity edges and add value to the pedestrian experience along the boardwalk. Public furniture is designed as public art elements with a robust nature and spatial uses are cleverly combined without utilities dominating focus.

PLACE 设计集团负责卡特斯码头景观顾问工作。该案是现代简约的城市设计的一个极好的例子，是布里斯班市最新的河滨项目之一。

景观色彩语言和几何与 HPA 的建筑互为补充，有时候对比为更广泛的区域形成特征。一个凉亭作为人行道连接河面的延伸浮桥通往电影院和餐馆。

河滨游泳池的设计是无限边缘的，为木板路上行人的体验增加价值。公共家具被设计成公共艺术元素，巧妙结合健全的自然和空间使用，而不必使用主要焦点。

Halcyon Waters

宁静水域

Landscape Architect / PLACE Design Group Pty Ltd.
Client / Halcyon Management
Location / Gold Coast, Queensland, Australia
Photographer / PLACE Design Group Pty Ltd.

The 27 hectare site is located on Hope Island, the Gold Coast's North Shore. About 40% of the site is developed for residential purposes. The development incorporates a Leisure Centre, recreational facilities and 227 detached houses based on zero-lot design principles, which emphasize individuality, privacy, controlled solar aspect and usable outdoor space. Stage 1 consisting of 100 homes is now complete and Stage 2 construction is underway.

The Leisure Centre provides a focal point for the community and incorporates a swimming pool, tennis court, cinema, gymnasium, workshop, along with a secret garden and garden studio. A dedicated private open space which we have named celebration park provides a wonderful opportunity to observe and absorb the natural environment.

Lifestyle always comes at a price, but Halcyon Waters delivers lifestyle at an affordable cost. Affordability is often confused with cheapness. At

Halcyon Waters affordability is aligned with value and value for money. Houses sell for about 60%~70% of the value of conventional house resales in the area, and this forms the basis of the marketing model.

Through the highest standard of design principles and innovation, along with market awareness and responsiveness, Halcyon Waters is helping to pioneer the new world of seniors living.

这片27公顷的小区位于黄金海岸北岸的希望岛，约40%被开发为住宅用地。该项目包括休闲中心、娱乐设施和基于零红线设计原则的227套独立房屋，强调了个性、隐私、控制太阳能和利用室外空间。1期的100套住宅已完成，2期工程正在进行中。

休闲中心是社区的一个焦点，包括游泳池、网球场、电影院、体育馆、工作室以及一个秘密花园和花园工作室。一个我们称之为庆祝公园的专用私人开放空间，提供了观察和融入自然环境的一个绝好机会。

优质的生活方式总是要付出高价格的，但 Halcyon Waters 却能以可负担的合理价格提供优质的生活方式。人们经常将合理和便宜混淆，在 Halcyon Water，可购性和价值以及金钱价值对齐。房屋售价约为该地区普通房屋转售价的 60%~70%，这形成了营销模式的基础。

通过最高标准的设计原则和创新，连同市场认知和反应一起，Halcyon Waters 协助开拓老年人生活的新世界。

Kunming Horti-Expo Eco-Communities

昆明园博生态社区

Landscape Architect / SWA Group
Client / Yunnan Horti-Expo Xingyun Real Estate Co., Ltd.
Location / Kunming, China
Area / 2,590,000 m²
Photographer / Tom Fox

SWA provided master planning and landscape architectural services with the goal of developing an urban center that simultaneously restored ecological balance through reforestation and watershed planning. Sustainable technologies treat and reuse wastewater, reduce peak flows associated with introduced urban storm-water runoff, orient buildings to reduce heating and cooling loads, use passive solar collection to reduce energy use, and cluster development in order to preserve and restore forest and hydrological corridors.

SWA also took a land-based approach to establish a series of highly integrated communities on the appropriate topography to avoid the inherent ecological degradation of extensive grading. Housing, retail, education, health services, and communications were also organized within walking distance of each other, offering residence access to goods, services, and a strong open-space system without having to rely on their automobiles.

SWA为这个项目提供总体规划和景观建设，在构建出一个新城市中心的同时，通过重新造林和流域规划来重塑生态平衡。设计上运用可持续技术处理和再利用废水，减少因城市降雨排水而带来的峰流，关注建筑本身以减少其供暖和冷却负荷，使用被动式太阳能蓄积系统以降低能源使用，采用集群发展模式来保护和再造树林和水文长廊。

SWA以土地为基础，在适宜的地势上建立起一系列高度相融的社区，以避免大范围的固有生态退化。住宅、零售、健康服务和通信等都设立在步行范围圈内，为住户提供商品、服务和成熟的户外空间系统，使得住户们不再需要依赖于汽车代步。

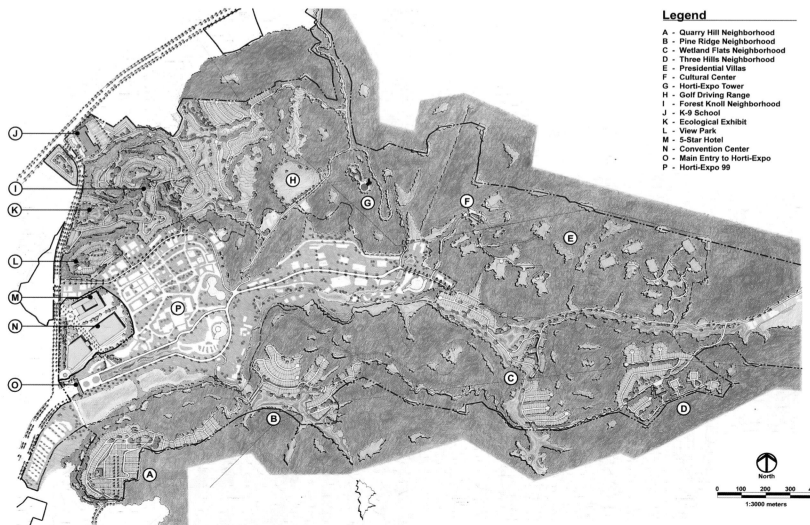

Legend

A - Quarry Hill Neighborhood
B - Pine Ridge Neighborhood
C - Wetland Flats Neighborhood
D - Three Hills Neighborhood
E - Presidential Villas
F - Cultural Center
G - Horti-Expo Tower
H - Golf Driving Range
I - Forest Knoll Neighborhood
J - K-9 School
K - Ecological Exhibit
L - View Park
M - 5-Star Hotel
N - Convention Center
O - Main Entry to Horti-Expo
P - Horti-Expo 99

Private Residence
私人住宅

French Residence

Frech 住宅

Landscape Architect / Mariane Wheatley-Miller
Location / City of Syracuse New York
Area / 4,046.85 m² garden with a further 16,187.43 m² woods
Photographer / Charles Wright

The concept of this project was to create a landscape for entertaining and enjoyment for the family and their two teenage children and friends. The design converted a steeply graded slope in the woods into terraces to allow for level outdoor entertainment facilities ensuring a gracious garden environment conducive to enjoyment and play.

Perched high on a hill, the site had magnificent views across the city and to wonderful sunsets. The designs needed large retaining walls to create two level terraces. The top terrace is comprised of a 'free form' swimming pool with a vanishing edge that hovers in the landscape. The vanishing edge drops 9ft to create a 'water wall' falling over natural stone and into a small pool on the lower lawn terrace.

The top terrace has gardens that surround the pool. A large stone fireplace acts as the center of this outdoor living room. We encouraged the client to rebuild their existing wooden deck and introduce a spiral metal staircase that freed up room on the terrace and improved circulation between the indoors and outdoors.

Stone steps and gardens lead visitors from the top terrace to the front of the house at a level ten feet higher up.

The lower terrace comprises a grass lawn and steps to a woodland walk. This area is for sports and play as well as setting up a tent for outdoor functions.

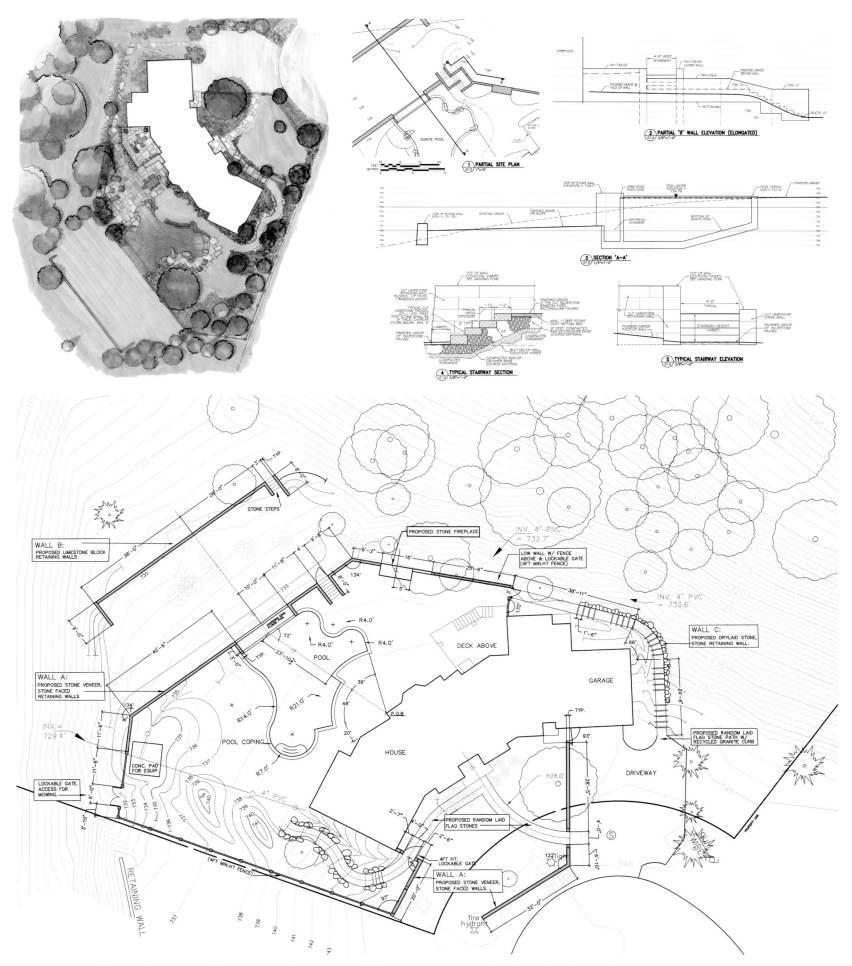

该案设计概念是为一个家庭和他们两个十几岁的孩子以及朋友们创建娱乐和享受风景的地方。设计将树林里的陡峭斜坡转变成梯级阳台，让同一层次的户外娱乐设施有一个可享受玩乐的优雅园林环境。

坐落在高高的山顶上，该处可俯瞰城市的壮丽景色和壮观的日落景色。该设计需要大型挡土墙以创建两层阳台。顶层阳台由一个边缘消失在景观里的"自由形状"游泳池组成，消失的边缘降低2.7米，在天然石头上创建一面"水墙"，落入较低处草坪阳台的小水池里。

顶层阳台由花园环绕，花园环绕游泳池和一座大石头壁炉，壁炉是这个户外客厅的中心。我们建议客户重建现有的木甲板和引入一个螺旋金属楼梯，连接阳台上的房间和增进室内外之间的循环。

石阶和花园引领游客从顶层阳台到3米高的房子的前面。

低层阳台由草坪和木板通道组成。这个区域是为运动和娱乐以及设立户外帐篷的。

LeKander Residence

利坎德之家

Landscape Architect / Mariane Wheatley-Miller
Architects / Grater Architects
Client / Mr. and Mrs. Dan LeKander
Location / Wellesley Island, St. Lawrence River, New York
Photographer / Charles Wright

The project included seven terraces in all, moving from the driveway and parking area down to the St. Lawrence Seaway. Three hard paved stone terraces with walls and steps leading down to the River are in the middle of the pedestrian path. Planting beds are on all three terraces as well as the final terrace comprising lawns and plant beds.

The middle terrace is comprised of a seating area, outdoor cooking facilities and a vine covered dining area under a timber pergola. The gardens that surround the seating areas are colorful year round and also include a 'cutting garden' filled with perennials providing a colorful summer display as well as a source for flowers placed inside the home.

The timber deck at the river's edge has seating areas and access to the boathouse and docks.

The priority of the project was to safely walk through seven level changes, but also to maximize the usage of these small areas and encourage outdoor living environments. Outdoor spaces are conducive to the frequent entertainment for friends & clientele of the owners.

The Existing house is on a small site — only 108 ft. wide by 93 ft. It is close to both neighboring houses where the community encourages public access across properties yet the design manages to provide a necessary sense of privacy for the client.

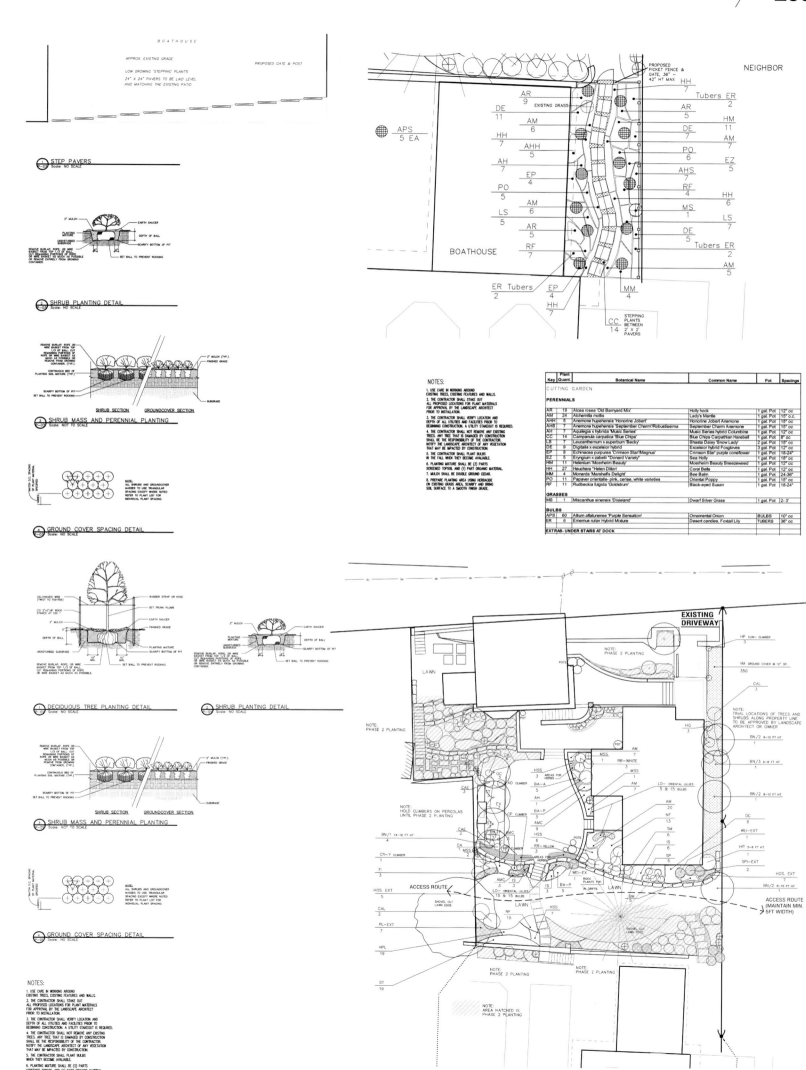

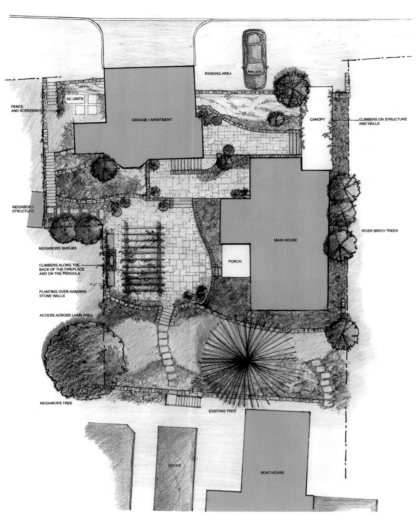

该案总共包括七个阶梯式露台，从车道和停车场往下一直到圣劳伦斯河畔。墙壁和台阶往下到河畔的三个硬铺面露台，都在人行道的中间。三个露台都有花坛并且最后的露台由花坛和草坪构成。

中层露台由座位区、户外烹饪设施和爬满藤蔓的木质藤架下的用餐区构成。座位区周围的花园全年都多姿多彩，还包括一个种满了多年生植物的"插枝花园"，这些植物在夏天百花争艳，是家里插花的花源。河畔的木甲板有座位区

并且可以通向船坞和码头。

该案的重中之重是安全地行走在这七个层次的变换中，同时也最大限度地使用这些小地区，优化户外生活环境。室外空间有利于业主朋友与客户频繁的娱乐。

现有房子只有 33 米长 28 米宽，靠近两个邻近的房屋，那里是邻里之间的公用空间，而设计则为客户提供必要的隐私感。

James Street Garden

詹姆斯大街花园

Landscape Architect / A J Miller Landscape Architecture
Client / Mr. and Mrs. Tony Miller the designers
Location / City of Syracuse, New York, USA
Photographer / Charles Wright

The concept of this project was to create a garden that engages the style and era of the home (Arts and Craft's period). The original garden consisted of asphalt pathways, overgrown trees shrubs and lawn. The front lawn, trees and shrubs were removed except for large trees on the bank at the sides.

The property sits high above James Street and appears to those in the garden of being up in the tree canopy. It is completely private and not overlooked or viewed by James Street. This aspect helped with the design concept of creating a garden for viewing from the raised porches of which there are three. One porch for seating and dining, the other two for seating and viewing the garden. This concept would lead to the creation of parterres in a Arts & Crafts pattern. 900 Boxwood shrubs were laid out and planted to create a precise Boxwood Parterre. The Parterre consists of six main squares. The back two squares have 'Holly Cubes' planted in them, while the front four squares have 'Henry Hudson Roses' that are pale pink when in bud turning to white as it flowers. This is a heavily scented Rose and free from disease or pests, and very cold weather hardy.

There is a four foot wide sandstone fountain sitting in the garden's center. The fountain acts as a sound buffer and as a bird bath.

Along the side of the garden and house to the west is a 'Rose Garland' constructed of galvanized posts that support a large diameter nautical rope. Many types of Roses, Clematis, Golden Hops and Dutchmans Pipe are planted with Oriental Poppies, Salvia and Lady's Mantle, along and below the garland.

该案理念是建造一个类似居家时代风格的花园（艺术和工艺时期），花园由沥青路、树木、灌木以及草坪构成。除了侧边堤岸的大树，前面的草坪、树木和灌木都被移除了。

花园坐落于杰姆斯街的高处，以树冠层的高度出现，是完全私隐的，在杰姆斯街大街无法俯瞰或者看到花园里的景色。这也有助于创建一个观景花园的设计理念，花园的高层门廊有三个：其中一个提供座位和进餐区，另外两个提供座位区观赏花园。这一概念将创建一个精致的工艺品图案。900棵黄杨灌木创建了一个精确的黄杨木花坛,花坛由六个方格组成,后面的两个方格栽种着"冬青方块"，而其他四个方格则是"亨利哈德逊玫瑰"，这种玫瑰在花蕾期是粉

红色的,当开放时则变成白色,是一种香味浓郁的玫瑰,不怕病虫害,而且非常耐寒。

花园的中心有一个四英尺宽的砂岩喷泉,可以减弱道路交通噪音,小鸟们还可以在喷泉里洗个澡。

沿着花园和房子往西,一个镀锌"玫瑰花环"柱支撑着航海粗绳。这里栽种了各种玫瑰、铁线莲、金色啤酒花和马兜铃属植物,花环周围和下面则种植了鬼罂粟、鼠尾草和羽衣草。

Garden 15

花园 15

Landscape Architect / BRUTO Landscape architecture
Location / Bled, Slovenia
Area / 4,000 m²
Photographer / Miran Kambič

This minimalist planned garden follows the sleek architecture of the residential building. The garden is designed in four units that differ throughout each season providing great attraction with their characteristic colors.

这个极简抽象艺术的花园跟时髦的住宅建筑相符。花园包括四个部分，随着季节变化，以它们标志性的颜色吸引着人们。

The western part represents spring or violet color with Japanese cherry tree and crocuses, the northern part represents summer or white color with roses and meadow blossoms, the southern part represents autumn or red color with decorative grass and Virginia creeper, the eastern part represents winter or green color with conifers.

西边以日本樱花树和番红花代表春天或者浅紫色；北边以玫瑰和草甸花代表夏天或白色；南边以装饰性草坪和五叶葡匐植被代表秋天或红色；东边以针叶树代表冬天或绿色。

Going Green

走向绿色

Landscape Architect / Rick Eckersley of Eckersley Garden Architecture
Location / Mornington Peninsula, Victoria, Australia
Photographer / Rick Eckersley

Climate change is having a real impact on the way people are approaching garden making in Australia. Our traditional approach of copying northern European gardens is failing as our country faces protracted drought, flood and fire. There needs to be a change in attitude toward gardens if they are to be successful in the future.

This garden is an experimental attempt to create a fully sustainable garden. The primary motivation behind this garden is to change public attitude to sustainable, drought tolerant, native gardens. There is a long-standing prejudice against the use of native plants — probably born of the 1970's when many bad native gardens were put in place. But horticulture has moved on since and there are many exciting, colorful, drought tolerant plants in the market place and they can be used in place of traditional exotic plants.

The swimming pool also follows sustainable design principles. It is a 'Natural Pool' laid out in a traditional rectangular design and built with traditional materials. What it does not have is any of the chemicals in the water that are traditionally used to keep water clean and healthy. Rather, it relies on a natural filtration system which runs the water through the roots of native water plants growing alongside the swimming area. The pool provides habitat wildlife and is like jumping into a crystal clear freshwater billabong.

在澳大利亚，气候变化已真正影响到人们设计花园的方式。传统照抄北欧园林设计已经行不通了，因为这个国家长期面临着干旱、洪涝和火灾。如果想要园林设计在将来成功的话，我们对园林设计的看法必须改变。

这座花园是为了创造完全可持续性花园的一次实验性尝试，主要动机是改变公众对可持续、耐旱、原生花园的态度。人们对采用原生植物的偏见由来已久，可能是源自20世纪70年代建造的许多不良的原生植物花园。但是，由于市场上有许多极具创意的、多姿多彩的耐旱植物，这些耐旱植物可以用来取代传统外来植物，园艺已经取得很大的进步了。

游泳池也遵循可持续发展的设计原则。这是一个用传统材料建造的传统矩形"天然游泳池"，水里没有任何用来保持水质清洁的化学物质，相反，它通过原生水植物生长在游泳区边缘的根，依赖于自然过滤系统。游泳池形成了野生物种的栖息地，就像跳进清澈见底的淡水洼地中一样。

Leafy Entertainer

多叶表演者

Landscape Architect / Myles Broad of Eckersley Garden Architecture
Location / Melbourne
Photographer / Sarah Appleford

Well-facilitated multipurpose spaces, interlaced and layered with selected plantings are appealing. Spaces where the plants envelop and entwine create the experience of living in the garden.

The main concept with the design of this project was to do just that. If the success of a space is how often it gets used, then this design is very successful as it gets a lot of use.

本案配置齐全的多功能空间，交错和层叠的精挑细选的植物非常吸引眼球，植物覆盖和盘绕的空间创造了住在花园里的体验。

该案的主要设计理念就是这样。如果以使用频率来评价一个空间设计是否成功的话，那么这个项目的设计就是非常成功的，因为其使用频率非常高。

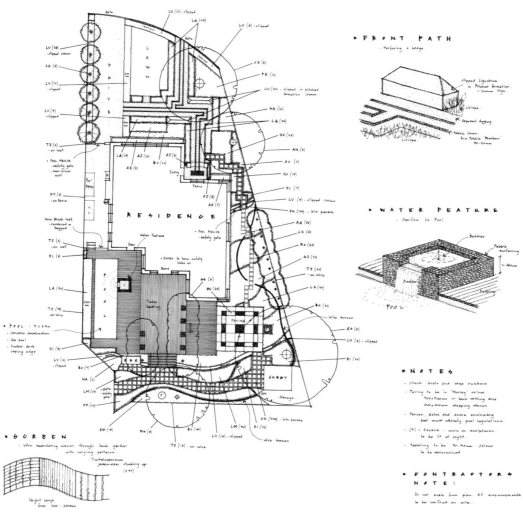

Kendall Residence

肯德尔住宅

Landscape Architect / Mariane Wheatley-Miller
Location / Potsdam, New York
Area / 8,093.71 m²
Photographer / Charles Wright

The creation of a wildlife habitat is aided by the placement of water in the garden and native plant materials are particularly useful for wildlife. Large lawn areas have been converted into wild flower meadows, flower gardens and fruit gardens and the remaining lawn areas are mowed less often. The overall site and garden design creates a natural and relaxed environment rich in color, texture and scent.

A lilac bluestone walkway leads one through gardens to the front door of the main house or turning into the private bluestone terrace. On this main terrace, a granite boulder that has been created into a water feature was installed. The water bubbles out through the top and disappears into river washed pebbles. Trees, shrubs and plants, surround the water feature and the composition allows for birds to drink and bathe in the fountain.

A second garden next to the main house named the 'courtyard' sits in between the garage, covered walkway and front garden. Stepping-stones lead one through this garden down some steps to the rear lawn area facing the river.

A private patio garden was designed for guests at the Cottage garden. Removing selected trees and shrubs enhanced the view to the river. The patio's stone is large natural flagstone with plants planted in-between the joints like wild Viola, Sedum and low growing Thyme.

To provide a sense of privacy for the neighbors adjacent to the compound we planted some hedging and trees. Hemlocks, Maples, Lilacs, Roses, Oakleaf Hydrangeas, and Witch-hazel were the species chosen.

该案设计通过花园里水的设置以及特别适合野外的原生植物来表现。大草坪区已变成了野花草地、花园和果园，并且剩余的草坪区则经常修剪。整个房屋和园林设计创造了一个色彩缤纷、质地丰富、香味芬芳的自然、轻松的环境。

一条丁香青石小径穿过花园通向住宅的正门或者通往私人青石阳台。这个主阳台上放置了一个创建水景的花岗岩巨石。水从巨石顶部冒出，消失在河水冲刷过的鹅卵石中。树木、灌木和植物，环绕着水景，这种构成让鸟类可以在喷泉里饮水和洗澡。

在住宅旁边的另一个花园就是"庭院"，位于车库、走道以及前花园之间。台阶石径穿过这个花园通向面对河流的后草坪区。

在小屋花园里设计了一个为访客准备的私人庭院，移除了选定的树木和灌木后，增强了河边的景色。庭院里的石头是大型天然石板，缝隙之间种有野生三色堇、景天属植物和低矮的百里香。

为了给住宅毗邻之处提供隐私感，我们种植了一些树篱和树木，如铁杉、枫树、丁香花、玫瑰、绣球花和金缕梅。

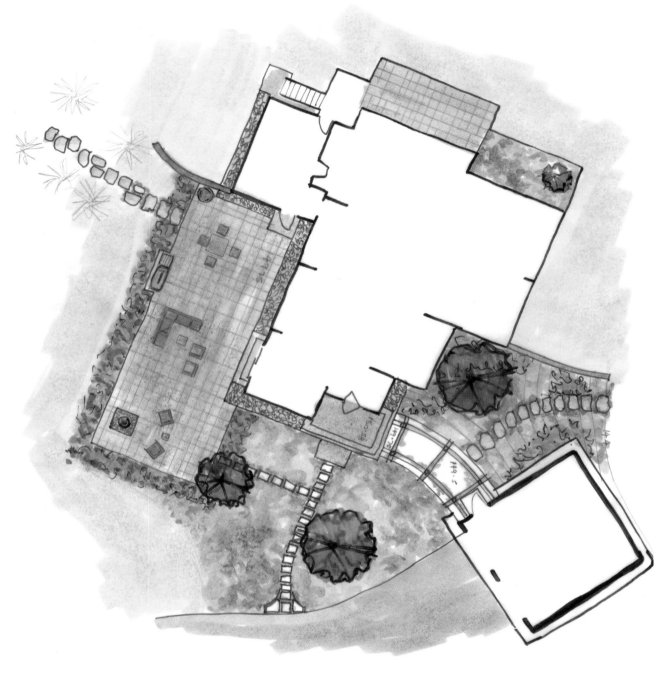

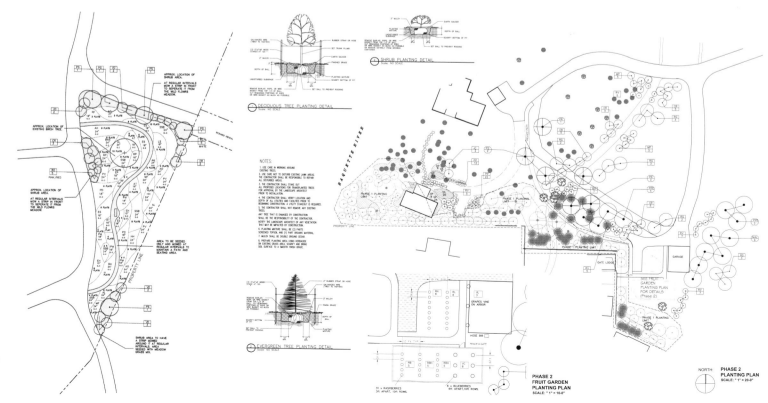

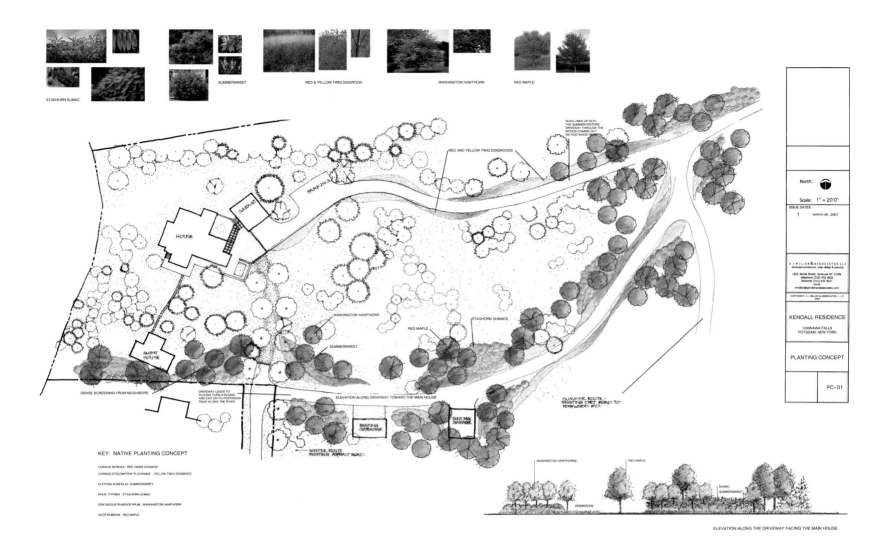

SITE LAYOUT & PLANTING CONCEPT

KENDALL RESIDENCE
HANNAWA FALLS, POTSDAM, NEW YORK

Glazer Residence

格莱泽之家

Landscape Architect / A J Miller Landscape Architecture
Client / Mr. and Mrs. Kevin Glazer
Location / Rochester, New York
Photographer / Charles Wright

The design was created as a garden reminiscent of the house's style. The client wanted a garden in a style similar to that they had seen in Europe. They requested a garden for entertaining and enjoyment for the family, especially their four young children and their friends.

We created a large brick terrace for outdoor dining and seating, this area was edged with large limestone troughs and planted out every season with lavish, dramatic, colorful displays. Pathways lead from here to the "Great Lawn" or under an archway planted out with many climbers and under planted with scented perennials. A vista begins from the terrace through to the formal "Parterre" enclosed in low Boxwood hedging. At the end of this vista there is a large limestone planter on a plinth with bench seating.

A second formal garden was created to the side of the house next to a raised terrace of the Living room. Here we designed a "Rose & Peony Garden" edged in Boxwood hedging and framed from the street with a row of mature columnar horn beams.

The "Great Lawn" was created by removing some large overgrown trees. With this tree canopy gone we leveled the lawn and re-laid new sod. Surrounding the lawn on one side we created a "Woodland Border" of trees and shrubs and shade perennials, some specimen trees and shrubs were used for distinction and color.

该设计让人联想起花园式住宅风格。客户想要一个跟他们在欧洲看到的风格相似的花园,一个供一家人,尤其是他们的四个孩子和朋友们娱乐和享受的花园。

我们为户外餐厅和座位创建了一个大型砖结构露台,这个空间镶有型石灰岩槽,种植了大量引人注目的季节性植物。从这里开始的小径到达"大草坪"或者是种植了许多攀爬植物和芳香的多年生植物于拱门下。

房子侧边的阳台旁边是另外一座花园，在这里，我们设计了一个黄杨木篱笆里围绕的"玫瑰＆牡丹园"，以一排柱状角梁框起了街景。

通过移除一些大型丛生树木，形成了"大草坪"，树冠移除后，我们修剪了草坪，重铺了新草皮。我们在周围草坪的一侧创建了一片"林地边界"，由树木、灌木和茂盛的多年生植物构成，为了区别和上色，我们采用了一些标本树木和灌木。

Private Garden – Dalkey, Co. Dublin, Ireland

都柏林多基私家花园

Landcape Architect / Maximize Design
Designer / Maximilian Kemper
Location / Dublin, Ireland

The gardens of this classic house from the 19th century were extremely overgrown and unmanaged for a long time.

The brief was to bring back the grandeur of the properties former golden days by opening up the front garden and restoring the "grand entrance", a driveway that gives approaching guests different vistas of the house and the front gardens. This sweeping entrance was also emphasized by positioning the main steps in the centre of the sandstone wall. Leading up from the steps is a small paved area which the family uses to greet their guests.

The back gardens, slope steeply upwards from the rear of the house, so tiered retaining walls were required to give enough space for a decent, family size patio. All walls in the design were built using Irish sandstone from Donegal for continuity and sustainability purposes.

Set at the top of the first flight of steps is a circular patio. It is a completive place where the last rays of sun of the day can be enjoyed under the magnificent presence of the old Sumac tree.

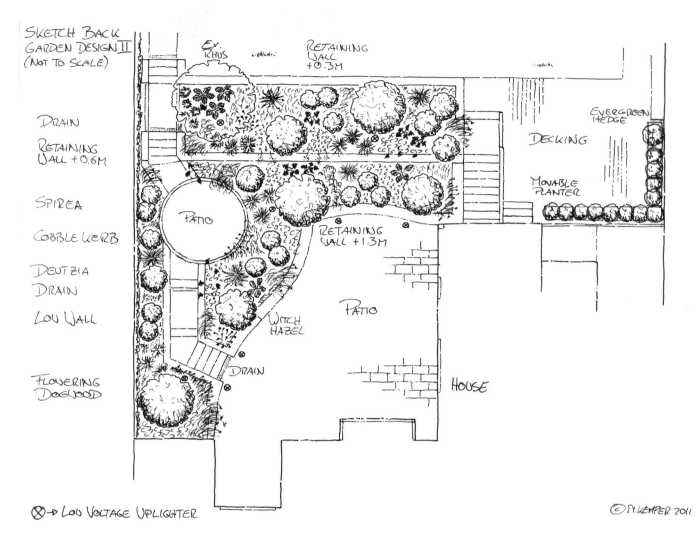

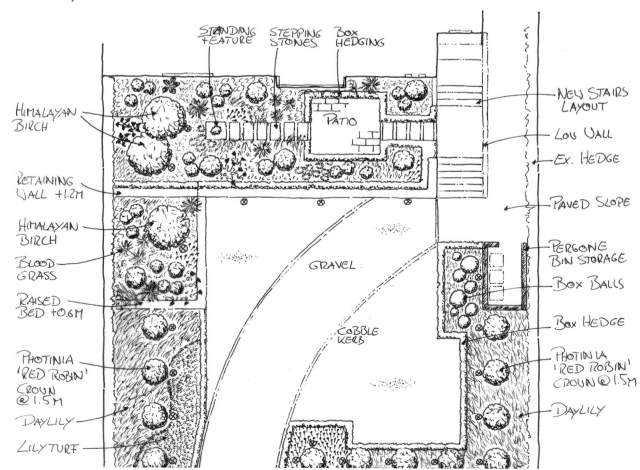

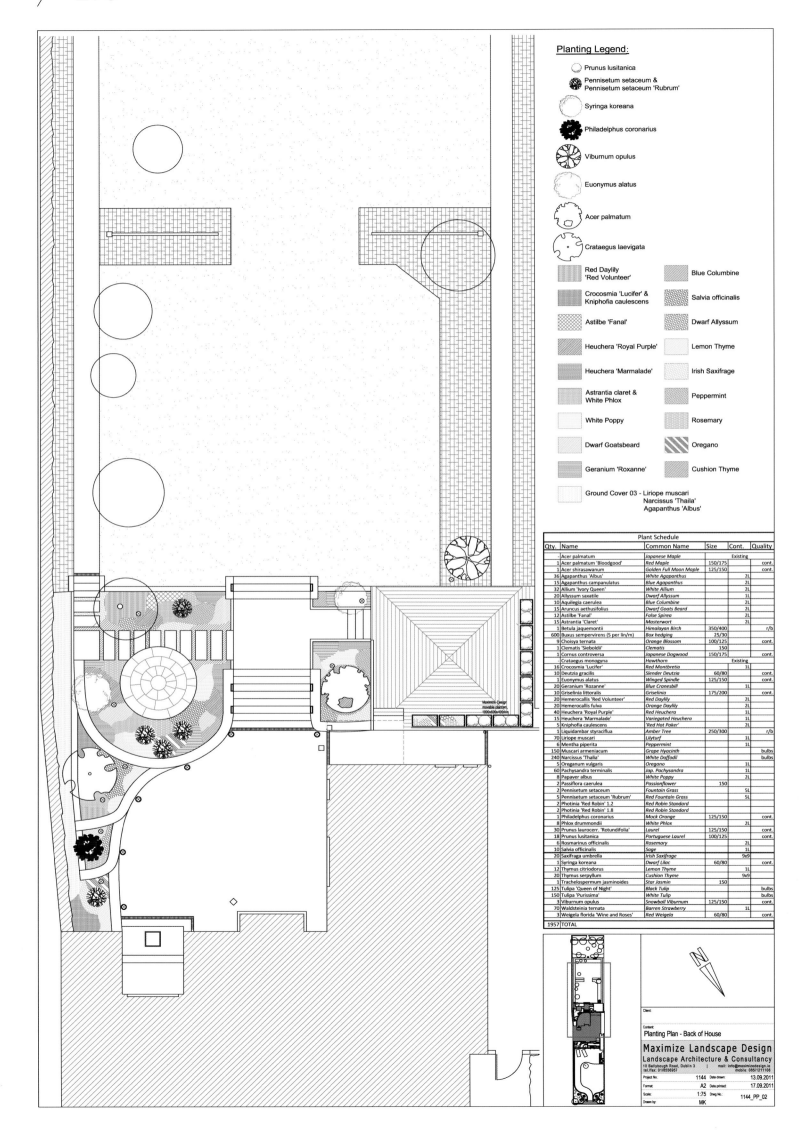

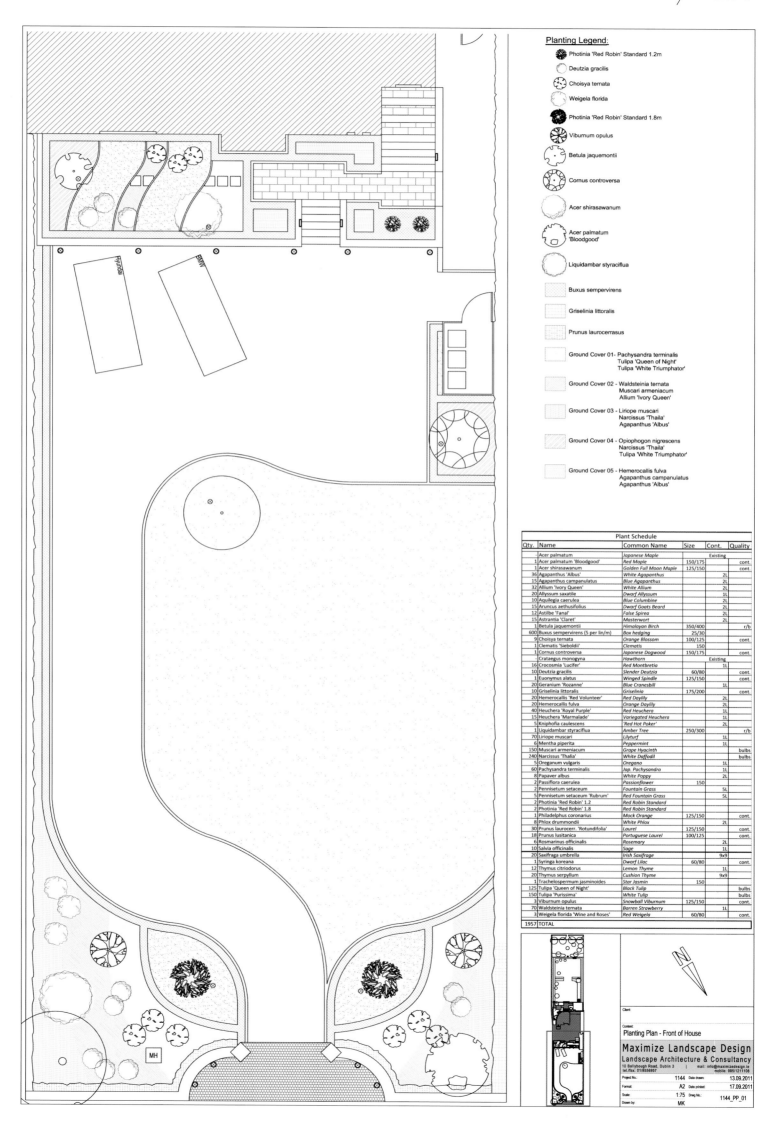

这座 19 世纪建成的房子周围的花园早已杂草丛生并且已很长一段时间没有打理过了。

该设计是通过完全打开前花园和恢复"宏伟入口"（一条让来访客人看到房子和花园不同远景的道路），再现原来黄金时代的辉煌。砂岩墙壁中央的主台阶的位置也突显了这包罗万象的入口，沿着主台阶往上走是一小铺面路，是这家人用来欢迎客人的。

后花园沿着房子后面大坡度地向上，所以分层的挡土墙要给出足够的空间，创建一个合适的家庭面积的阳台。为了连续性和可持续性的目的，设计的所有墙壁都采用多尼戈尔爱尔兰砂岩。

阶梯第一段的顶部设置了一个圆形阳台，是一个完全可以在伟岸的老漆树下享受一天里太阳最后一缕光线的地方。

Roxbury Renovation

罗克斯伯里之家翻新

Landscape Architect / Mark Rios, FAIA, FASLA, principal, and John Fishback of Rios Clementi Hale Studios
Location / Los Angeles, CA, USA
Area / 929.03 m²
Photographer / Dominique Vorillon

Rios Clementi Hale Studios created a dramatic residential showplace for long-time clients. The original, 1930s Paul Williams-type house was in good condition at purchase, but the new owners wanted to make changes to meet their lifestyle and taste. The new Roxbury home would blend the traditional aspects of the existing house with the owners' affection for 50's and 60's modernism.

While the owners were accustomed to living in the huge, sprawling space of the previous home, they wanted each room at Roxbury to have its own identity. The designers looked to the existing elements of each room to solve the program concerns of the clients. From the brick exterior, with custom-designed sconces, into the entry, the home exudes a gentle manner.

Existing landscape elements of pool and pond remain, but are replanted to give greater texture and uniformity through the practice of bunching large amounts of like plants together. Around the house, the plant palette is simplified to complement the architecture, while a big-leaf garden meanders through and around the pool and ponds. A small surprise garden is created with a geometric pattern of black and white stones dotted by succulents.

工作室为长期客户创造了一栋非凡的住宅式剧院。原住宅是 20 世纪 30 年代保罗威廉姆斯式的，购买时房屋条件都良好，但新业主想做出改变，以适应他们的生活和品位。新的罗克斯伯里之家将融合现有房屋的传统方面与业主偏爱的 50 年代和 60 年代现代主义风格。

业主一家已习惯生活在原房子的巨大宽敞空间中，所以他们新家的每个房间都有各自的特色。设计师在看过每个房间现有的元素后解决了客户的问题。从砖墙外部与定制的烛台到入口处，都散发着高雅的礼仪。

保留了游泳池和池塘现有的景观，但移植了现有植物，通过大量相似植物捆在一起形成更大的纹理和均匀性。房子周围，植物色调简化为跟建筑互补，一个大叶花园蜿蜒流过泳池和池塘。多肉植物点缀的黑白石头的几何图形创建了一座小小的惊喜花园。

Mediterranean Garden

地中海花园

Landscape Architect / Rosmarin Landscape Design
Location / Cleveland Heights, Ohio, USA
Photographer / Ann Rosmarin

This garden offers a blend of formal and informal elements, perennials and unusual tropical throughout, koi and lily ponds, a vegetable and cutting garden.

这座花园处处都融入了形式和非形式的元素、常见和不常见的热带风情，有锦鲤和百合花池塘，是一座如剪裁般的绿色花园。

Garden of a villa in Dahlem

达勒姆花园别墅

Landscape Architect / Silvia Glaßer und Udo Dagenbach
Location / Dahlem, Berlin, Germany
Photographer / Udo Dagenbach

The basic design principle for the home garden was to shield the surrounding neighbouring areas and structures from the garden and the structuring of the garden itself in diverse areas, which respectively enables the use for all family members.

The crucial idea of the reorganization of the garden was implemented using linear and selective relements from plants (Fagus sylvatica, Carpinus betulus, Buxus sempervirens) and stone (limestone walls, limestone cuboids, gabion walls of limestone, terrace slabs of limestone, basalt stone paving).

The repetition of the stone elements and the play of colours in the plants composed from a mixture of red-leaved trees, reinforces the impression of the garden by the appearance linear tree plantings and hornbeam walls along the property border.

In general, only white and blue purple flowering shrubs, perennials and bulb plants were used, thus stimulating a greater depth effect of the landscape, through the selective use of adequate flower colours.

The interaction of the recurring materials and plant structures in different dimensions results in a spatial expanse.

家居花园设计的基本理念是在不同区域让周围邻近区域和结构不受花园和花园本身结构的影响，从而使得所有家庭成员都可使用。

花园重组的重要思想是使用线性和选择性的元素，通过植物（水青冈、鹅耳枥、黄杨）和石头（石灰岩墙壁，石灰岩立方体、石灰岩笼壁、石灰岩厚板、

玄武岩石铺面）来实现的。

　　石料和植物色彩，如红叶树的重复，通过花园边界上的线性树木和鹅耳枥木墙增强了花园印象。

　　在一般情况下，只种植白、蓝紫色的开花灌木以及多年生球茎植物，通过选用适当的花的颜色，从而刺激景观更纵深的效果。

　　在不同规模中反复出现的材料和植物结构的相互作用形成了广阔的空间。

Squire Creek Residence

斯格尔溪住宅

Landscape Architect / Jeffrey Carbo Landscape Architects
Location / Lincoln Parish, Louisiana, USA
Photographer / Chipper Hatter, Hatter Photographics; Louisiana Helicam; Jeffrey Carbo Landscape Architects

Our role involved site master planning, design review of exterior living spaces attached to the home, preparation of construction documents, construction administration, and project observation for all hardscape and landscape features. The residence had been "aggressively" cleared by a previous owner, and was supposed to turn "lemons to lemonade" by using cleared spaces for needed open areas for play.

The use of native aggregates for drive surfaces, stone that emulated a native rock found nearby, storm water retention as a rain garden, and the pool as a minimal yet functional and elegant garden element contributes to our design intent. The early comprehensive involvement and collaboration with the owner and architect reaped many benefits with the completed work and allowed us to test and implement many intricate design details.

Preserved wooded areas became backgrounds for designed spaces, with walking paths integrated throughout. Native grasses and wood fern planted generously and in bold gestures contrast with the simple lines of lawn. The limited plant palette also responded to the client's maintenance concerns, yet gave us the opportunity to create striking and dynamic spaces that could be memorable in a minimal way.

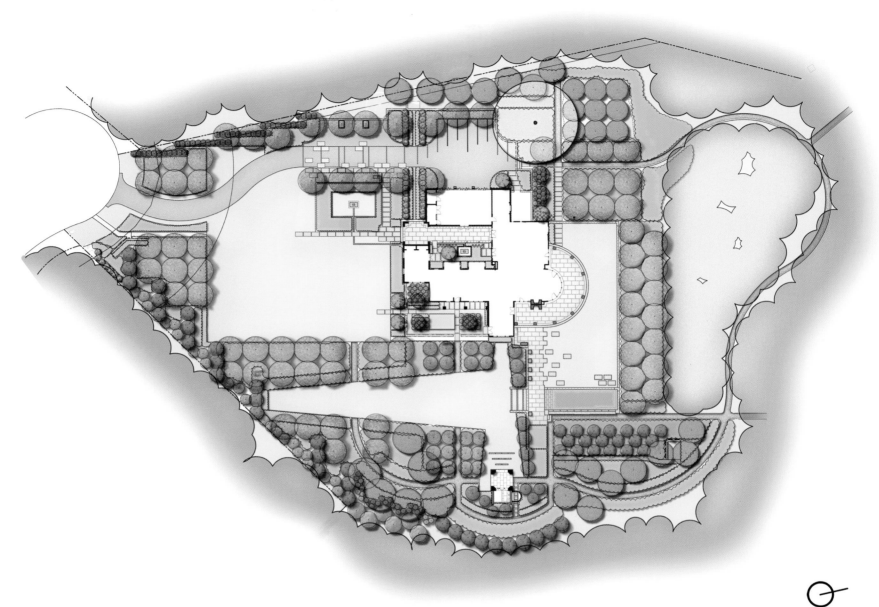

我们参与了该案的总体规划，跟住宅相连的室外空间的设计审查、施工文件准备、施工管理以及景观的项目观察。这座住宅被前主人"激烈地"扫荡过，为了营造玩耍所需的户外空间，需要通过明亮的空间将"柠檬变成柠檬汽水"。

本案设计构成采用了以下元素：车道表面的本土混凝料、模拟附近岩石的石头、作为雨水花园的雨水收集器，以及一个小游泳池，一个极小但功能齐全的优美的花园池。早期的全面参与和业主及建筑师的合作极大地帮助了该案的

完成，让我们能够测试和实施许多复杂的设计细节。

保留的树林区成为了设计空间的背景，与步行道融为一体。大量种植的原生草和木蕨类植物以醒目的姿态与草坪简单线条形成对比。有限的植物色调回应了客户关于维护方面的忧虑，然而也让我们有机会以一种最简单的方式创造值得纪念的引人注目的动态空间。

House on the Hills

山上的房子

Landscape Architect / Beatriz Santiago
Architect / Architectare (Flavia Quintanilha + Rodrigo Fernandes) + Pedro Quintanilha
Location / Rio de Janeiro, Brazil
Area / 614.23 m²
Photographer / Leonardo Finotti

Despite being in a proportionately large lot, the building has the shape defined by the small triangle of its buildable area on the entrance of the lot. The whole building area is embedded in this triangle, with the exception of the deck, which was designed as permeable construction as allowed.

Because it is very exposed, the facade towards the street was designed to give the impression that it is all closed, with exception of the library, which has the bookshelf. The graphic texture of this façade represents the integration of the architectural project with the land. From the soil, a well defined form arises. It is the land being transformed into architecture.

On this strong base, another volume is lightly settled. This part represents the designing architecture. In this volume, the louver's rationality is interrupted by the library, which is characterized by the disordered wood graphism, representing a storehouse of ideas.

Because it is a cold place, the façade towards the inside of the land is made of glass, receiving direct sun during the afternoon and warming the house for the night. Furthermore, the glasses provide the necessary input of light during the day, since the facade towards the street is all closed. To increase the interior light, a skylight above the stairs was designed.

The entire 1st floor is designed to ensure full interaction among the different uses and spaces of the house. For this, the doors can be completely open, transforming the living room into a veranda and the deck becomes an extension of the internal space, annulling the barriers between the swimming pool and outside kitchen.

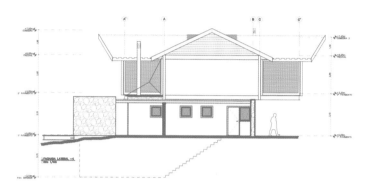

facade um
no escale

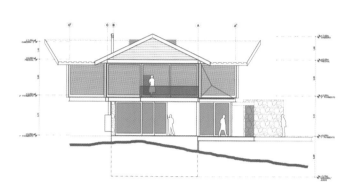

facade three
no escale

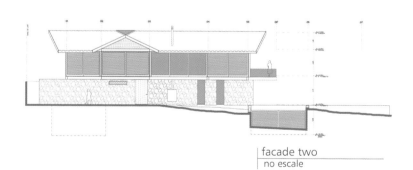

facade two
no escale

facade four
no escale

section a-b
no escale

section c-d
no escale

尽管占地面积很大，但建筑的形状由其地块入口处建筑区的小三角形来决定。总建筑面积楔入到这个三角地带中，甲板除外，甲板是作为渗透结构而存在的。

因为位置很暴露，面向街道的外观的设计给人的印象是这房子是全封闭的，书斋例外，因为有书架。房子外观的图形质地表现了建筑土地一体化。泥土中出现了一个良好设计的形状，这是土地正在被转化为建筑的过程。

在坚实的地基上，轻轻放置着另一个建筑体。这部分就是设计的建筑，在这个建筑物里，百叶窗的理性排列被书斋中断，无序的木质图形构成了书斋的特点，体现了思想宝库。

因天气寒冷，嵌入土里的房屋外层是玻璃幕墙，吸收午后阳光在晚上让房子变暖。此外，由于面对街道的一面完全封闭，玻璃幕墙在白天提供必要的阳光输入，为了增加室内光线，在楼梯上方设计了天窗。

整个一楼的设计确保了不同功能和房子之间的充分互动。为此，房门可以完全开放，将客厅转变为阳台和甲板，成为内部空间的延伸，清除了游泳池和户外厨房之间的障碍。

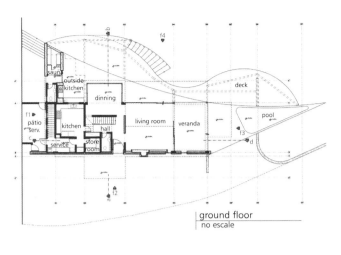

ground floor
no escale

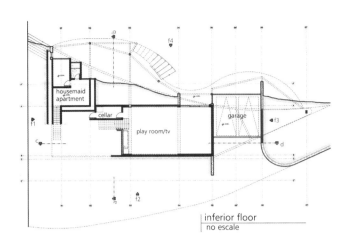

inferior floor
no escale

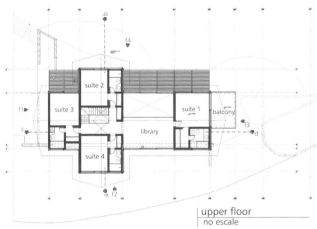

upper floor
no escale

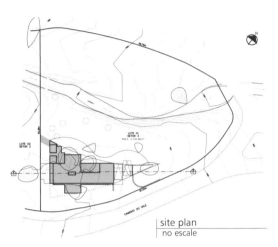

site plan
no escale

Between Predictable & Unexpected

预见与意外之间

Landscape Architect / Terragram
Project Team / Vladimir Sitta, Terragram
Architect / Diego Balagna
Location / Sydney, Australia
Photographer / Vladimir Sitta

To clearly distinguish new and old, the building volumes are separated by insertion of two tiny gardens. The client, an art collector, appreciated our tendency to seek sculptural qualities in designed landscape. Unusual construction techniques were used to construct green mounds (including shredded paper as landscape paper mache). A theatrical atmosphere is evoked by incorporation of artificial mist. The living room window frames a very narrow space that is notoriously difficult to treat. Small leafed plants survive on charred and sculpted trunks, which were specifically selected by designers in the Blue Mountains, Australia. We personally hacked, sliced and charred the trunks, before they were transported on site. The installation and spatial arrangement was done directly on site. This project is another example demonstrating how important the personal presence of the designer on site is. The ground floor garden contains refurbished and enlarged pool and simple palette of evergreen plants that form a green carpet under predominantly existing trees. Earthworks unraveled sandstone outcrops that lend the area stronger character and help to anchor building elements in the site.

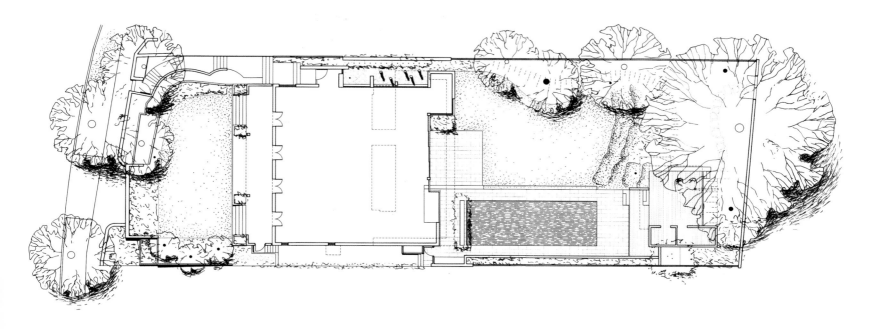

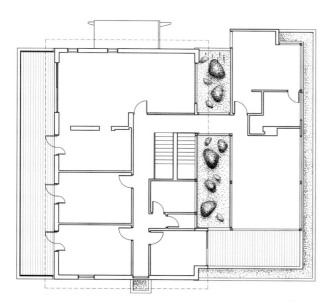

明确区分新旧，该建筑体由两个小花园分开。业主是一位艺术品收藏家，欣赏我们倾向于寻求景观雕塑般的设计品质。特殊施工技术被用来建造绿丘（包括作为景观用纸的碎纸片）。人造雾团诱发了一个剧场般的氛围。客厅的窗户框起了一个非常狭窄的空间，不易处理。小叶植物存活在烧焦的雕刻树干上，树干是设计师在澳大利亚的蓝山山脉中特别选定的，亲自砍伐，切断和烧焦树干，然后运往现场，安装和空间布局是直接在现场进行的。这个项目是诠释在施工过程中设计师的存在是多么重要的另外一个例子。一楼花园包括翻新和扩建的池子以及常绿植物，在树下形成了一片绿色的地毯。土木工事解开了砂岩岩层，协助固定建筑。

Moltz Landscape

墨尔茨景观

Landscape Architect / Ibarra Rosano Design Architects
Location / Tucson, Arizona, USA
Area / 120.77 m²
Photographer / Bill Timmerman

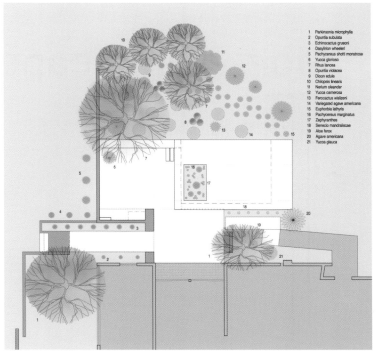

1. entry hall
2. kitchen
3. pantry
4. powder
5. laundry
6. t.v. room
7. dining
8. living
9. office
10. west bedroom
11. west bath
12. west courtyard
13. east hall
14. east bedroom
15. east bath
16. shower courtyard
17. master bedroom
18. master bath/closet
19. outdoor fireplace
20. dining patio
21. pool
22. trash enclosure
23. south deck
24. carport
25. entry plaza

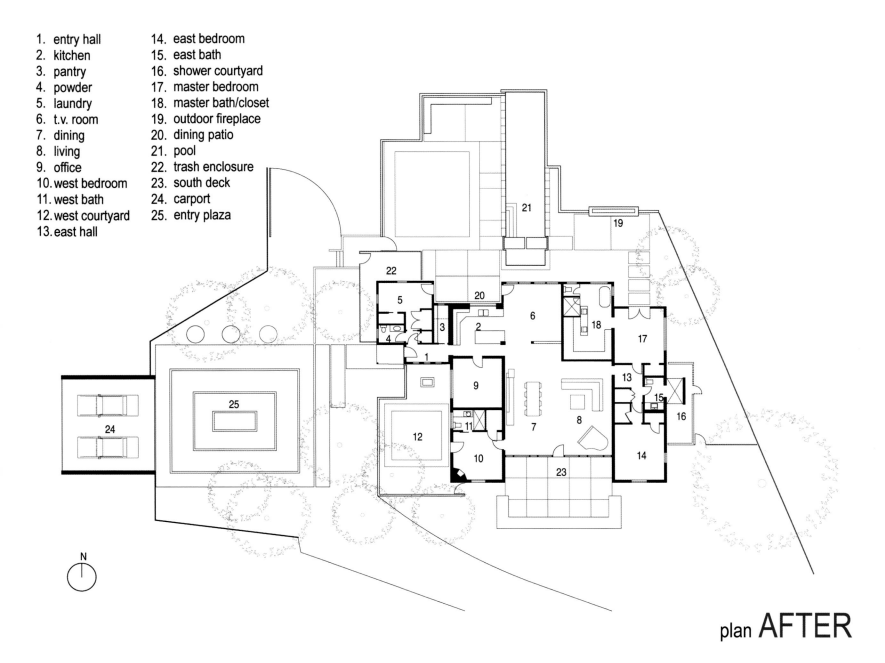

plan AFTER

1. entry
2. kitchen
3. pantry
4. powder
5. laundry
6. t.v. room
7. dining
8. living
9. office
10. west bedroom
11. west bath
12. east bedroom
13. east bath
14. master bedroom
15. master bath
16. pool
17. carport
18. sunroom
19. porch

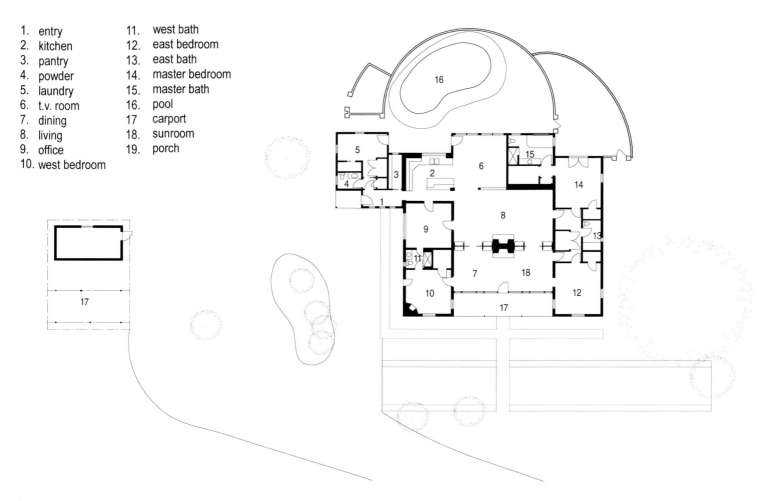

plan BEFORE

On the east side of the house mature trees volunteered in complete randomness. So an ordering system was needed to give logic to their presence and harvest their shade. Also, the land sloped awkwardly at the house, washing mud onto a brick patio outside the entry door. Therefore our intervention had to also correct this problem.

To cue a visitor toward the front door, which was not at all in the front but on the side, a bold marker was obligatory, but not so bold as to overpower the house.

A double-cantilevered, site-cast concrete wall is the visitor's first experience. The path begins beneath this gateless gate, which invites guests through a new masonry patio wall. This concrete block patio wall is like a stratified formation that plays horizontally against the sloping earth.

Once through this understated wall, the design unfolds to reveal a framework of hovering horizontal planes rendered in colored concrete. The stacked concrete slabs create planters, benches and outdoor space, which cantilever quietly above the desert floor. The owners experience is that of sailing on a stationary lawn chair, cool drink in hand, into the desert's horizon.

房子东边的大树是完全随机排列的,因此需要调整,给它们以逻辑性存在和形成荫蔽。同时,房子的地面奇怪地倾斜,将泥土冲到外面入口的砖头阳台处,因此,我们的设计必须同时解决这个问题。

为了给走向前门的访客提示(其实不是真正的前门而是房子侧面),需要设置一个加粗的提示牌,但又不能粗到超过房子的比例。

首先映入访客眼帘的是现浇双悬臂混凝土墙。道路从这无门之门的下面开始,客人需通过一面新的砖瓦阳台墙。这混凝土块阳台墙就像是一个分层的形式,在倾斜的地面维持着水平。

一旦通过这面朴素的墙壁,设计展现出在彩色混凝土中呈现的悬停水平面框架。堆叠的混凝土板形成了花盆、长椅和户外空间,在那里悬臂安静地悬在沙漠地板上。业主的体验是:冰爽饮料在手,在一个固定的草坪椅上滑行到沙漠的地平线。

GR House

GR 房子

Landscape Architect / Gil Fialho — Planejamento e Paisagismo Tropical
Architect / Bernardes + Jacobsen Arquitetura
Location / São Sebastião, SP, Brazil
Area / 580 m²
Photographer / Leonardo Finotti

The implantation covers an area of 580 m² with the proposal of a construction that contains an extensive program. Therefore, the design comes with the necessities of many openings for illumination and ventilation in a narrow land between two other constructions in the condominium. With a few plans of façade, the distribution of the plant is in three levels with the floor that contains the service area embedded half level, the main floor (living room and two bedrooms) half above of the arrival level.

This concept made possible the integration of the external and the internal. The construction was implanted in two blocks having defined two empty areas. One is in the front of the construction with the garden of entrance and parking of cars. And other, the main one, with a great private internal garden contained by the blocks of the construction. The fluid circulation of people is realized through catwalks, with no limits, no walls. What it provides a playful environment. Central garden as axle of the house converges of all the spaces. The central garden is also designed with magical elements in accordance with the Japanese concept of engawa.

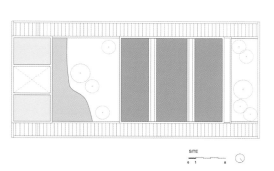
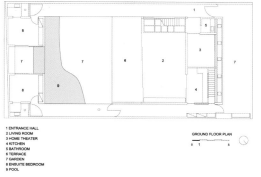
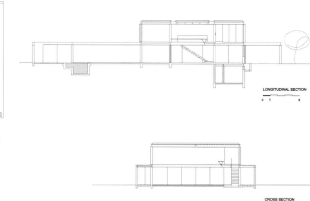
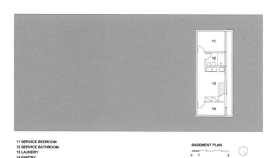
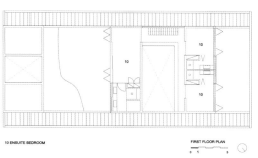

该案占地 580 平方米，包括一个大量设计规划的建筑结构。因此，在复式住宅其他两个结构之间的狭窄土地上，需要设计许多用于照明和通风的开口。住宅外层设计上，植物的分布和楼层一起分三个层次，该楼层包含嵌在半层的服务区以及高于入口层一半的主楼层（客厅和两个卧室）。

这个理念让室内外融为一体成为可能。建筑设置在定义空区的两地块上，一块是住宅的前半部分，是入口花园和停车场；另一块是房子的主结构，有一个被建筑块包围的极隐蔽的内部花园。人们可以通过步行小道在院内畅游，没有墙壁的阻隔，花园所提供的是一个有趣的环境。中央花园作为房子的轴心汇聚了所有空间，中央花园的设计，也根据日本的缘侧（engawa）设计理念采用了奇妙的元素进行设计。

/ 334　Private Residence　私人住宅

JN House

JN 房子

Landscape / Fernando Chacel/ Burle Marx e Cia
Architect / Bernardes + Jacobsen Arquitetura
Location / Itaipava — RJ
Area / 503,471.42 m²
Photographer / Leonardo Finotti

Since the house is always full and has frequent visitors, it was necessary to create a program cautiously thought to create the perfect wellness and relaxation centre.

A topographical approach was developed to sculpt the land and to integrate the architecture with nature. Consequently, the construction was distributed through the land, being basically a single storey house characterised by different independent blocks. Concrete structures and garden slabs bind into the topography and others subtly stand out through a mesh of wood in the landscape, created through a succession of structural frames.

The access to the house is through a paved pathway leading to a suspended volume. This creates a front porch with a stunning view of the Rocky Mountains in the background.

From the tennis pavilion you can see a stone wall, which generates privacy and hides the service areas: gourmet kitchen, lavatory, storage and pantry. The gourmet area is covered by a wooden structure supported by thin round metal pillars.

The garden slabs form a ceiling for some blocks of the house, so indoor comfort is maintained for visitors. The fireplace and interior gardens help regulate the temperature. The existence of large glass panels and skylights lets in natural light.

由于这座房子常常人满为患，有许多访客，因此有必要打造一个精心设计的完美休闲中心。

该案设计采用了地形手法，塑造土地，并使建筑与自然融为一体。因此，结构分布到土地上，是一个由不同独立块组成的单层住宅。混凝土结构和花园板结合到地形和其他结构上，通过该地景观木网巧妙地突显特色，形成一连串的结构框架。

通向房子的是铺面路，引向悬浮的建筑体块，从而创建了一个可欣赏到洛矶山脉绝妙景色的前廊。

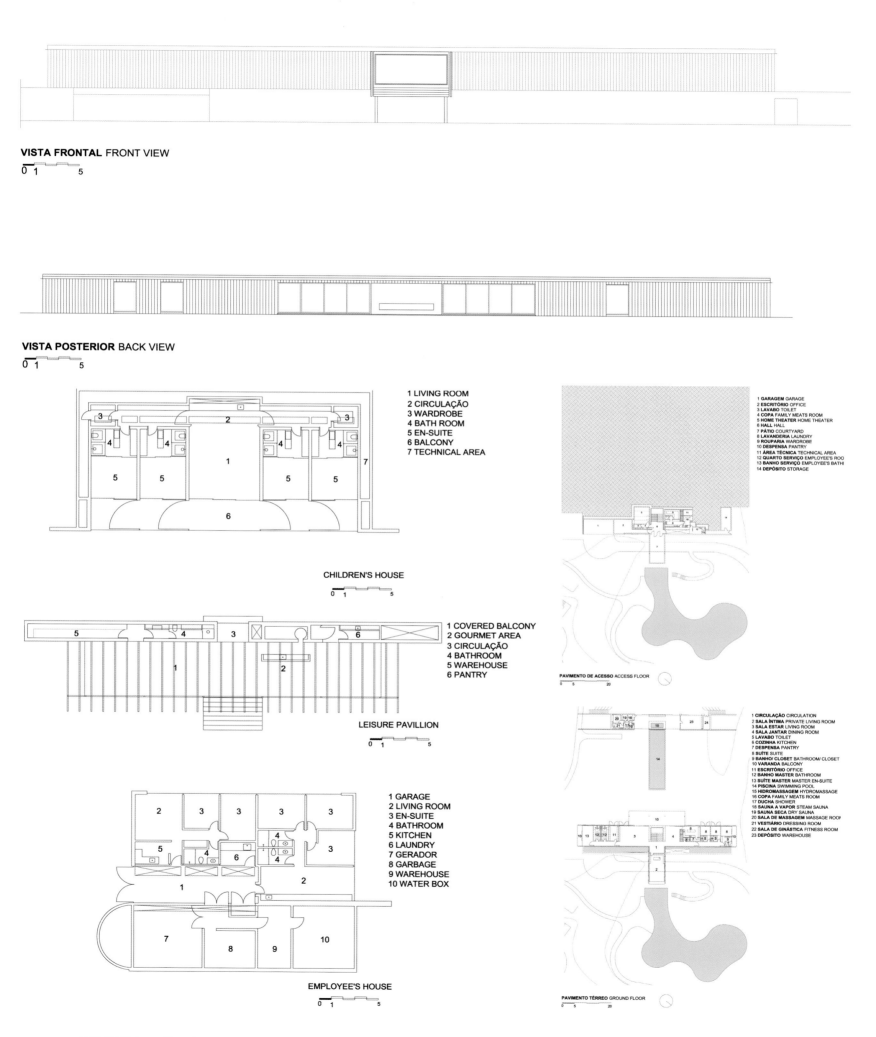

从网球馆看去，可以看到一面石墙，它可以提供隐秘空间和隐藏服务区域：美食厨房、洗手间、仓库和储藏室。美食区由薄圆形金属柱支撑的木质结构覆盖。

花园板形成了房屋的部分天花板，所以访客在室内依然可以感受到舒适度。壁炉和室内花园协助调节温度，大型玻璃窗和天窗让自然光进入室内空间。

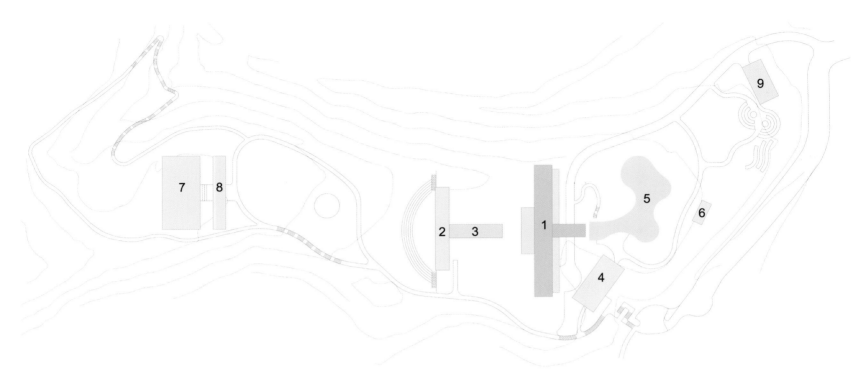

IMPLANTAÇÃO SITE
0 10 50

1 **CASA PRINCIPAL** MAIN HOUSE
2 **SPA** SPA
3 **PISCINA** SWIMMING POOL
4 **CASA DOS FILHOS** CHILDREN HOUSE
5 **LAGO** LAKE
6 **CANIL** DOG'S ROOM
7 **QUADRA DE TENIS** TENNIS COURT
8 **PAVILHÃO DE LAZER** LEISURE PAVILLION
9 **CASA DO CASEIRO** EMPLOYEE'S HOUSE

AMB House

AMB 房子

Landscape Architect / Jundu Paisagismo
Architect / Bernardes + Jacobsen Arquitetura
Location / Guarujá, SP, Brazil
Area / 1,310 m²
Photographer / Leonardo Finotti

The AMB House is situated on the coast of São Paulo, Guaruja City in the middle of the Atlantic Forest. From the street you can only see one of the three floors of the house because the terrain has accentuated slopes that give different views of an almost untouched natural landscape.

The most common situation, where the rooms are upstairs and social rooms are downstairs, was reversed on the design of this residence. On the entrance of the house we have a hall that serves as a mezzanine overlooking the double ceiling living room with proximally 6m height and wood frame glass windows with a view looking at the swimming pool and the forest. In the lateral of the hall, the balcony outside is surrounded by large wooden bench.

In the intermediate floor are the social areas: the living room and dining room, which join the outdoor kitchen, outdoor deck and infinity swimming pool.

Downstairs is the intimate area with five suites. It is this strategy of reversing the usual array of social and intimate area that makes the rooms, even overlooking the sea, have the privacy afforded by the trees that are at that level.

The house can be called the balcony house with large glass panels that allow visual contact with the surrounding areas of the residence and the natural landscape of the region. The windows of the rooms have wooden Cumarú shutters. This wood is also present on the deck of the balconies and floors of rooms. In the living room two bamboo plants sprout from the middle of the floor, bringing the forest into the house.

This is how the house shows its relationship with the local landscape, barely visible through the dense forest in the access road, but grows and can be seen on the other side along with the look of a coastline and a stunning tropical forest.

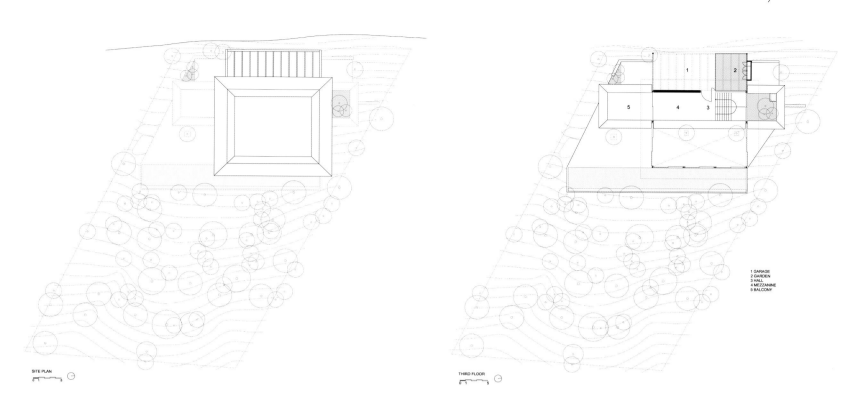

SITE PLAN

THIRD FLOOR

1 GARAGE
2 GARDEN
3 HALL
4 MEZZANINE
5 BALCONY

AMB 房子坐落在大西洋沿岸森林中部的瓜鲁雅市的圣保罗海岸线上。从街上你只能看到房子三层中的一层，由于地形加剧了斜坡坡度的缘故，这条斜坡形成了原始自然景观的不同景色。

住宅一般楼上是住宅区域，楼下是公共空间，而这个住宅设计是刚好相反的。房子的入口处有一个大厅，作为一个阁楼，俯瞰着双天花板的约 6 米高的客厅和可看到游泳池和森林的木框玻璃窗。在大厅侧面，巨大的木长凳环绕着户外阳台。

中间层是公共领域：客厅和饭厅，连接户外厨房、室外甲板和宽敞的游泳池。

楼下是五间套房的私密空间。正是这种策略逆转了一般的交际和生活区，使房间甚至可以俯瞰大海，而且拥有由在同等水平面上的树木提供的私密空间。

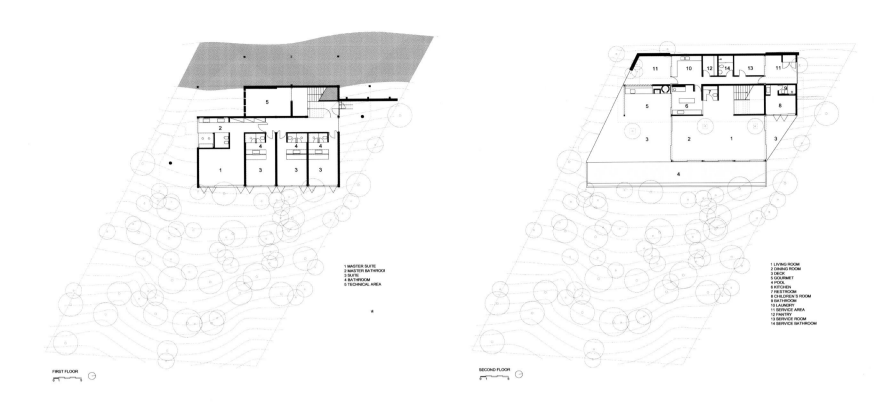

这座房子可以称为阳台房子,有大型的玻璃板,与住宅周边地区和该处的自然景观直接进行视觉接触。房间窗户是巴西柚木百叶窗,这种木材也出现在阳台房间地板上。在客厅里,两种竹类植物从中间的地板上破板而出,将森林带进了房子。

这就是房子怎样跟当地的景观联系在一起的,在路边茂密森林里是几乎看不见房子的,但随着海岸线和一个迷人的热带森林一起,也渐渐可以在另外一端看到。

Private Residence 私人住宅

Remanso de Las Condes, Casa C

雷曼索德拉斯康德斯，C 住宅

Landscape Architect / Karla Aliaga
Client / Inmobiliaria DICAL
Location / Santiago, Chile
Photographer / Italo Arriaza

The project is characterised by the extensive use of concrete, with a contemporary look both for the living space as well as the exterior distribution. This material, together with the wooden elements, propose a clear though subtle contrast with the natural components of the patio. This material together with a marked timber pose yet subtle contrast with the natural elements of the court, noting instances, denoting paths through the undergrowth which in turn accompanies these circulations bounded by trails cross the decline, which allows a much fuller contemplation walking environment, recognition of this through colors, textures and others like the contrast of materials, finishing this in the lower part of the pool area which merges both the eye and the other travels through different topographic levels that the project has.

该案特色是大量混凝土的使用，与现代外貌的生活空间及户外分布。混凝土和木质材料，与庭院的大自然景色形成了一个明显而又微妙的对比，指向穿过灌木丛的通道，通道相应地伴随着坡下行的小径分界的循环，通过颜色、纹理和其他如材料对比形成一个更完整的冥思步行环境，该案的设计以位于较低处的游泳池结束，通过不同地形高度将视觉景色和其他路径融入其中。

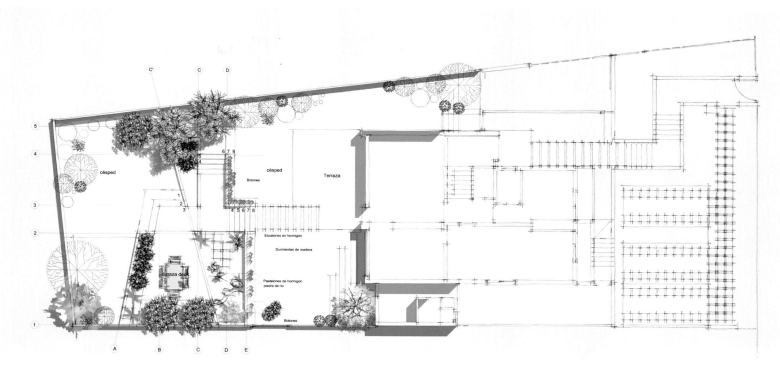
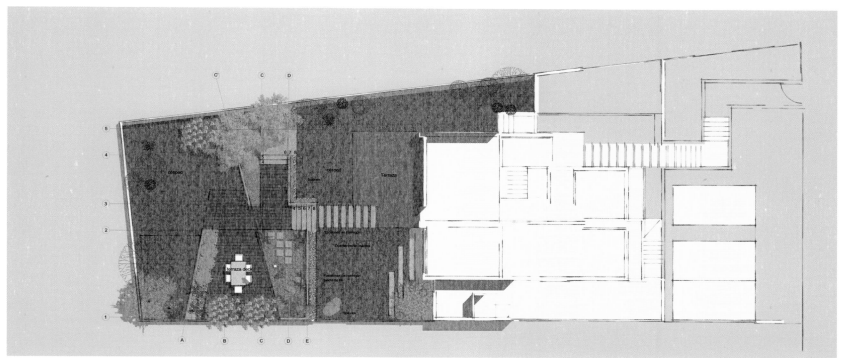

San Carlos 1

圣卡洛斯 1

Landscape Architect / Karla Aliaga
Location / Santiago, Chile
Photographer / Italo Arriaza

Addressing a need to fill the outside spaces, just as within a home, we configured a number of spaces in the courtyard of this house. Here takes place the development of these spaces, to shape the areas in which the family lives outdoors.

Places like the barbecue and terrace made of wood creating a light screen, but with timber sleepers highlighting their own height, with steps placed according to what the terrain requires, highlight the intention of creating spaces for actual use and grant value to the landscape as a result of this. It proposes an order and harmony as a result of an intention of use of outdoor spaces, and by the utilization of materials such as brick and concrete, place the resident almost in an interior living mode.

要满足户外空间的需求，就像在家里面一样，我们为这座房子的庭院配置了许多空间。为了创建让家庭可以居住在户外的空间，因此才有了这些空间设计。

像木制的烧烤场和露台这些地方形成的光线屏风，但跟突显自身高度的枕木以及按照地形需求设置的阶梯一起，强调了实际空间的使用意图，从而赋予了景观价值。该设计提倡整齐与和谐，出于使用户外空间的意图，通过材料如砖和混凝土的应用，使居住者处于室内居住模式中。

San Carlos 2

圣卡洛斯 2

Landscape Architect / Karla Aliaga
Location / Santiago, Chile
Photographyer / Italo Arriaza

When performing both an architectural and landscape project, we are influenced either directly or tacitly by a freely generated intent from the inclination of an owner or the designer for a particular form or in this case for a given material. It is the use of stone which is being proposed here, as we transform it into the tool that will bring into harmony the elements that give character to the backyard of a house.

The stones on the ground and the patio, subtly highlight a visitor's tour on this surface, so as not to disrupt this order, and items such as stone pots, becoming part of the land, all arise from the original intent in simple, clear, harmony of each element with another.

当同时执行建筑和景观项目时，我们直接或间接受到某种自然形成的意图的影响，这种影响来自业主或者设计师对某种特定形式的偏好或是偏爱某种特定的材料。在此案中，这种材料便是石头，我们将其转换为工具，为构成房子后院特色的元素带去了和谐之感。

地面和露台的石头巧妙地在表面突显观光路线，以免扰乱这种秩序，其他物品的设计均来自最初的意图，以简单、明确、与其他各要素和谐一致的面貌出现，例如，成为土地一部分的石盆。

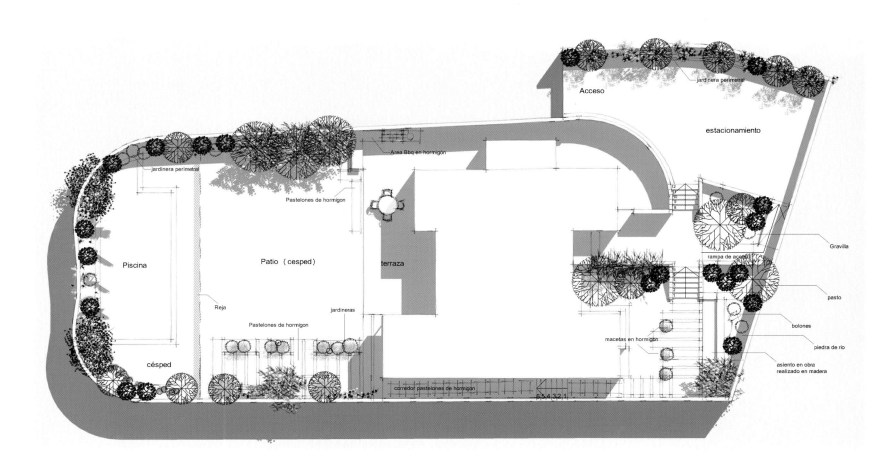

House in Ramat Hasharon

拉马特·哈莎伦的房子

Landscape Architect / Levy Chamizer & Mor Avidan
Location / Ramat Hasharon – Israel
Photographer / Shai Epstein

Our clients hoped their new home to feel like a resort. It was our challenge to fulfill that dream while accommodating the daily needs of a family with four lively children, all of which enjoy entertaining a lot of guests all year round.

The "L" shaped house was positioned on the plot so that it embraces the garden and pool. All rooms face the splendid outdoors. The layout also features a wide entry pathway that proceeds along a vertical garden, creating a facade covered by an entanglement of plants.

We created a house without windows: all openings are high and slim iron framed glass doors. The huge openings allow the merging of indoors and outdoors in order and maintain maximum eye contact between the public spaces on the ground floor. Even the upper level bathroom openings are doors leading to the balconies with their great white blossom views.

The 20 meters long swimming pool is aligned on the center axis of the master bed, allowing the owner, a professional athlete, to jump into the pool as soon as he wakes up in the morning.

An outdoor kitchen and bar were designed to accommodate large parties, taking advantage of the warm weather that allows outdoor cooking from April through November.

The architectural envelope is composed by a restrained pallet: a white structure, black iron glass doors, grey shingled roof tiles and the colors green and white taken straight from the lush garden.

我们的客户希望他们的新家像度假村一样。这是我们的挑战，实现这样的梦想的同时要满足一个有四个活泼的孩子的家庭的日常需求，而且全年都招待很多客人的家庭。

"L"形房子位于一个包含花园和游泳池的地块，所有房间都面向绚丽的户外风景。房子布局使得其出入也极其方便，道路沿着垂直花园，形成了一个植物覆盖的房子立面。

我们创建了一座没有窗户的房子:所有的开口都是高高的超薄铁框玻璃门,为了保持地面层公共空间之间最大的视觉接触,巨大的开口将室内外空间融合在一起。连楼上浴室的开口也是玻璃门,以白色的大花图案通向阳台。

20米长的游泳池与主人床的中心轴成一直线,让职业运动员的业主,可以在早晨醒来的时候能立刻跳进游泳池里。

户外厨房和酒吧是为了大型聚会而设计的,利用4月到11月温暖的天气,可以在户外烹饪。

建筑的外层是朴素的,白色结构、黑色铁玻璃门、灰色木瓦屋顶以及来自葱郁花园的绿色和白色构成了建筑的外层。

Narla

娜拉住宅

Landscape Architect / Peter Glass & Phil Cummins
Location / Pittwater, Sydney, Australia
Photographer / Helen Ward

This exciting pool, spa and landscape project in Sydney's Pittwater area were part of a major house and landscape renovation that converted a very average suburban house on one acre of land into a unique self-contained resort.

The design firm specializes in the design of swimming pools and outdoor spaces, and in particular in the seamless integration of internal and external areas. This can be clearly seen in the design of the outdoor lounging area, complete with fire-pit and surround seating, as well as in the design of the cabana and outdoor cooking areas adjacent to the pool, which all work very successfully.

The extensive landscaping surrounding the house and pool, including the tumbled granite driveway and the expansive lawn and garden areas, serve to further enhance the generous proportions of this property.

这一令人兴奋的游泳池、SPA和景观项目是悉尼碧水区一栋大房子和景观改造工程的一部分，景观改造是将一座位于一英亩地块上的普通郊区房子转变为独特的度假胜地。

该设计公司擅长游泳池和户外空间设计，特别是室内外空间的完美融合。这些都很容易通过该案的户外休息区以及火炉和环绕的座椅设计看出来，还有毗邻游泳池的小屋和户外烹饪区的设计，所有的工作非常成功。

房子和游泳池周围大量的景观，包括减速带花岗岩的车道以及宽阔的草坪和花园区，进一步提升了该房屋的比例。

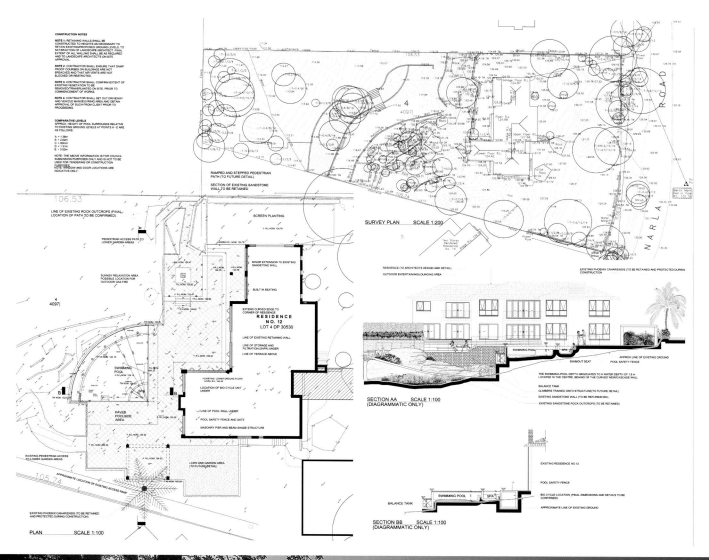

Fuleihan Residence

Fuleihan 住宅

Landscape Architect / Mariane Wheatley-Miller
Location / Syracuse, New York, USA
Area / 261.99 m²
Photographer / Charles Wright

The concept of this project was to create a patio garden that had a flavor of the Lebanon and Greece, both countries that our Client and their family visit often. The area is used frequently for "alfresco dining", swimming and entertaining with family and friends.

As the garden area is only a small space and overlooked by adjoining properties, we created some privacy with fencing yet created an open and bright feeling. The existing brick paving were removed and replaced with a light beige Travertine. The lighter color reflects the client's eclectic collection of souvenirs and colorful blue ceramic planters, the pool and many colorful plants.

该案设计理念是建造一个有黎巴嫩和希腊风情的阳台花园，因为业主和他们的朋友常去这两个国家。该阳台是经常使用的"露天餐厅"，是与家人朋友一起游泳和娱乐的地方。

由于花园面积很小，且被邻近住宅俯瞰的特性，我们用篱笆创建了一些隐私空间，而且是开放、明亮的感觉空间。现有的铺砖被移除，换上了浅米黄洞石。花园中布置了丰富多彩的蓝色陶瓷花盆和多姿多彩的植物。

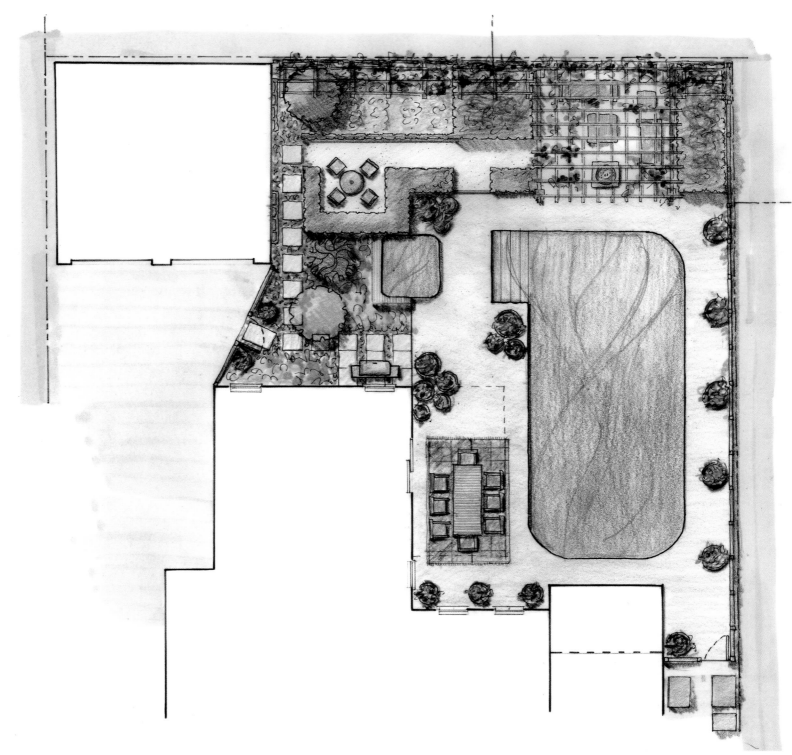

Private Residence "Water Canvas"

私人住宅"水上帆船"

Landscape Architect / Terragram
Project Team / Vladimir Sitta, Terragram
Architect / Chris Elliott Architects
Location / Bronte, Sydney, Australia
Photographer / Vladimir Sitta

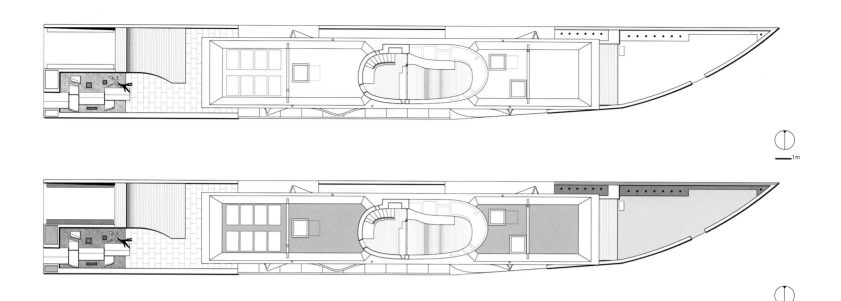

This tiny garden is the entry courtyard to a private residence by architect Chris Elliott, in Sydney's beach suburb of Bronte. Large slabs of Chinese grey-green granite are suspended over water, forming a path through a floating carpet of duckweed and water ferns. Stone was deliberately cut — some stone edges are split roughly, others sawn precisely. Small depressions carved in the stone, were marked directly in the factory by designers. These retain water after rainfall, while the granite path when wet, becomes a mirror to the sky. Two of the stone slabs have a horizontal slot to "release" water into the pond. Duckweed is usually considered to be the curse of fishponds covering the entire surface of water. This very shallow aquatic roof garden reverses the act of its removal. The courtyard relies on the presence of duckweed, leaving only

small sky windows — frames that are not in fixed positions, but rather could be frequently rearranged. These frames keep plants from invading the small geometric space. Exposure to sun transforms the color of plants from bright green to reddish brown, with shifting winds continually transforming and remixing this water canvas of plants. Small patches of different species give the surface an almost painterly appearance. Despite very shallow water, small native fish finds a safe haven under the floating green carpet of aquatic plants. The client (the architect of the house himself) and landscape architect, harvested aquatic plants from sites, where they were considered an unwanted nuisance.

这个小花园是建筑师克里斯·埃利奥特私人住宅的庭院，位于悉尼海滩郊区的勃朗特。大量的灰绿色中式花岗岩石板悬浮于水面，浮萍的漂浮地毯和水藻形成了一条小径。石块经过谨慎的切割处理，一些石块边缘大致劈开，另外一些石块边缘则是精确锯出来的。石块上雕刻的小凹槽是在工厂就直接做好的，这些凹槽可以储蓄降雨后的雨水，而当花岗岩小径被淋湿时，会变成反射天空的一面镜子。其中两块石板带有水平槽，可将雨水"释放"到池子里。浮萍通常被认为是水池的累赘之物，会覆盖着整个水面；这座很浅的水顶花园则反转了浮萍的这种扩张习性。庭院依靠着浮萍的存在，只留下小天窗，这个框架不是固定位置的，而是可以经常重排的，这些框架阻止植物入侵到小几何空间里。暴露在阳光下的浮萍颜色从亮绿色变成红棕色，和时常变换的风向一起不断转换和重组这艘植物水上帆船。不同物种的植物块形成了绘画般的外观。尽管水很浅，但是原生小鱼仍可以在漂浮水生植物的绿色地毯下找到一个避风港，业主（建筑师本人）和景观设计师，会捞掉一些他们认为造成滋扰的水生植物。

Winter Residence

冬日住宅

Landscape Architect / Ibarra Rosano Design Architects (Luis Ibarra and Teresa Rosano)
Location / Tucson, Arizona, USA
Area / 329.81 m²
Photography / Bill Timmerman

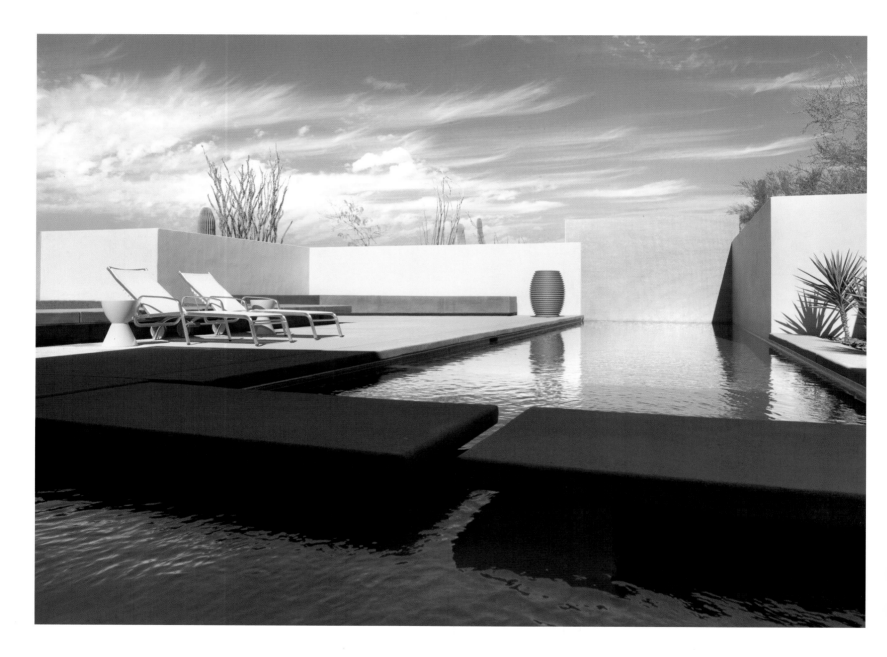

The client wanted to transform the dark, circulation-dominated rooms of their 1940s brick residence into luminous spaces with crisp detailing like in the boutique hotels and spas they had visited. Tackling one room after another, we opened the house to daylight and to a series of garden courtyards.

One key to transforming the house was to remove the fireplace that bisected the main living spaces. Once the fireplace was demolished, the opening for the chimney became a large skylight that now fills the living spaces with daylight. Applying the same principle, we removed the dividing wall between the master closet and bathroom to make a large, bright space for dressing and bathing. Outside, the area that had once been an awkward circular dirt driveway became a serene courtyard with a fountain, a single tree, and a horizontal slot opening that perfectly frames the city skyline--a favorite yoga and meditation spot.

The guest bathroom also connects with the garden. In this case, the glass-enclosed shower (the only additional square footage) extends out into a private courtyard. A pair of "floating" concrete bridges over the pool links the main deck to a master bedroom deck with its own outdoor fireplace. After over a year of living in the dust of remodeling, the house has become the owner's own private spa in the desert.

　　业主想要将这座 40 年代砖瓦房子中黑暗的房间变成明亮的空间，拥有像精品酒店和水疗中心那样的清晰的细节。在解决一个又一个房间后，我们为房子带去了日光和花园庭院。

　　改造房子的关键是移除平分了主要生活空间的壁炉。一旦壁炉被拆除，烟囱开口处就成为一个大天窗，阳光透过天窗填满了目前的居住空间。同样原理，我们拆除了主壁橱和浴室之间的分隔墙，形成一个更大、更明亮的空间，可以更衣和洗浴。户外曾有一条难以处理的脏兮兮的圆形车道，如今已成为一个宁

静庭院，一座喷泉，一棵树和一个横向槽的洞口，洞口完美地框住了城市天际线——最适宜做瑜伽和进行冥想。

客房的浴室也与花园连接。在这种情况下，封闭的玻璃淋浴（唯一的附加面积）延伸到私人庭院。水池上面一对"浮动"的混凝土桥连接主卧的主甲板和户外壁炉。生活在改建工程的灰尘中一年多以后，这座房子成为业主在沙漠中自己的私人水疗中心。

Contributors

AHBE Landscape Architects

AHBE Landscape Architects is an award-winning LA-based firm that translates revolutionary ideas into lasting environments, collectively offering extensive experience in the technical development of design aesthetics and constructability that are hallmarks of their work. As modern landscape architects in a postmodern world, AHBE sees the role of the landscape architect as providing the crucial link between ecology and economy, and striking a harmonious balance between nature and humanity. The firm's body of work includes a wide variety of project types and scales, spanning gardens and parks, medical facilities, recreation and athletic facilities, civic plazas, streetscapes and "green streets," mixed-use commercial developments, schools and institutions, transit oriented development, and others.

A J Miller Landscape Architecture PLLC

A J Miller Landscape Architecture PLLC is a landscape architecture firm founded on the basic assumption that good landscape design emerges from a strong understanding of how people use space coupled with the unique features of the specific space being designed.

Anthony Miller and Mariane Wheatley-Miller combine their design backgrounds into a team that is dedicated to the realization of high quality, well-crafted landscape designs. The firm employs a theoretical yet practical design process, always engaging the site and its clients/end users. In short, our core philosophy combines measured practicality with an inspired adventurous design spirit that both our clients and we can lay claim to.

Close attention to the bridge between concept and detail has made our firm popular with builders and contractors and has helped maintain good relations with the entire project team. Our firm offers complete landsape architectural services and prides itself on maintaining personal contact with all those engaged in the design project.

Architectare (Flavia Quintanilha + Rodrigo Fernandes)

Architectare was founded by Flavia Quintanilha and Rodrigo Fernandes.

Established in 1999, our goal is to design spaces that are not limited only by the practical function and shelter aesthetic, creating spaces that positively stimulate their users.

In 2001, received a Honorable Mention for The Residence Carvalho/Corção in the 1° Architecture Competition for Americas MNBA – Arqa 21 and had the same work displayed at the Buenos Aires International Biennial of Architecture.

Since its beginning it has designed more than 100 projects in several categories, such as residences, housing, hotels, commercial and institutional.

ARQUI-K Arquitectura+Paisaje

Arqui-K was founded by owner/chief designer Karla Aliaga in 2007, and is based in the Lo Barnechea district of Santiago, Chile. The studio specializes in the design and construction of landscape architecture projects for private and corporate clients, from small terraces and patios, to vast expanses in urbanization projects. About 2/3 of the clients are from the Santiago metropolitan area, and the rest from other regions of Chile. The design team is comprised of architects, designers, and ecologists. The focus of the work is getting to know and understand the client to generate a global view of how they live, work and relate, so as to extract the essence of their lifestyles and transfer it to warm spaces with special details which promote their family life.

Bernardes + Jacobsen Arquitetura

The office Bernardes+Jacobsen was created in 2001 fruit of the union between the architects Thiago Bernardes and Paulo Jacobsen. In 2005 Bernardo Jacobsen became partner of the office. Consequence of a history that crosses three generations, the office became one of the biggest architecture office in Brazil, with more than 1000 built projects including residences, offices, leisure, entertainment, institutional and furniture design. An architecture that seeks an equilibrium between interior and exterior, the plastic use of materials and the client's needs and wishes has been settled as the office's own language.

Nowadays, BJA is developing projects of architecture and urbanism in Brazil and abroad, although has been divided into two cells: Bernardes Architecture and Jacobsen Arquitecture, both with head offices in Rio de Janeiro and Sao Paulo and staffed with over 50 collaborators.

BRUTO Landscape architecture

BRUTO Landscape architecture is engaged in complex planning, where they put the same weight into landscape design and projects of space programs. They believe that, beside good design each given area needs a project of suitable space programs as well as purpose of use, if we want the area to be complex, colourful in terms of programs, useful and alive. So it is more important how it works than how it looks. They believe in the context of space, which can be manifested as an influence or as a guiding principle in project design.

For successful realisation of a project good project definitions are necessary, this exposes both the problems and the goals. Each task can be a challenge, should it be a garden, a park, the highway or natural environment.

They plan entirely, from project regulations to the smallest detail, which is why they undertake each task in an interdisciplinary way, by connecting the various experts. This is necessary for a qualitative and expert realisation of each individual project.

Eckersley Garden Architecture

Eckersley Garden Architecture is a boutique landscape design firm based in Melbourne that brings a revitalised approach to landscape creation from the renowned stable of Rick Eckersley. The business operates from two locations - the principle office in the Melbourne inner city suburb of Richmond, and a secondary office in Flinders, a rural coastal area of the Mornington Peninsula. Long time associates and recent partners Scott Leung and Myles Broad along with Kathryn Green join Rick to bring knowledge, innovation and passion to a changing industry. We also work closely with leading Australian architects.

Years of experience in design, sustainable garden principles, horticultural science and building construction gives Eckersley Garden Architecture the ability to take the art of garden making to new levels in a market place which values fresh ideas and requires a new, sustainable approach to gardens. Eckersley Garden Architecture receives commissions Australia wide and internationally. Our client base is as varied as our garden designs, ranging from small residential, to commercial multi residential, to country retreats.

Guangzhou Homy Landscape Co. Ltd.

Guangzhou Homy Landscape Co. Ltd., a design agency specializing in landscapes and responding to the needs of clients, is positioned at combining its top international design team and local personalized services.

Most of Homy's chief designers are the senior foreign designers who worked for a long time in some international and first-class institutions, such as Belt Collin Hong Kong, Belt Collin Thailand, Hong Kong ACLA, Pan Asia International and so on. They have been the chief designers of landmark works, like Guangzhou Chateau Star River, Favorview Palace, Shangri-La hotel and Ritz Carlton hotel, Hong Kong International Exhibition Center etc.

With its highly recognized design quality and considerate personalized services, Homy has become the long-term strategic partners of domestic top developers, such as China Merchants Property, China Overseas Company, RandF Properties, Hopson Development Holdings Limited, and CITICGroup etc.

A domestic mainstream client has commented, "We usually prefer Homy when it comes to the local institutions as we usually prefer EDAW when it comes to the international company ".

Ibarra Rosano Design Architects

Luis Ibarra and Teresa Rosano (AIA, LEED AP) are native Tucsonans and graduates of the University of Arizona College of Architecture. In 1999, they founded Ibarra Rosano Design Architects, and have since earned national recognition as one of Arizona's top design firms for their modern desert architecture. Their work has been published internationally and has received over fifty regional and national design awards. They were selected by Architecture Magazine as one of nine firms to represent their state in its issue on "the Arizona School", and their work was exhibited at the Scottsdale Museum of Contemporary Art in its architecture and design show featuring work from six firms based in the southwest region. They were honored with Residential Architect's "Rising Star" National Leadership Award in 2008 and were recently included among 50 international firms for their "ra50: Short List of Architects We Love" for their body of work to date.

Jeffrey Carbo Landscape Architects

From its beginning in 1995, JCLA has continued to refine its skills to become one of the South's leading landscape architecture firms in the United States of America. Under the direction and guidance of its founder, Jeffrey Carbo, FASLA, the firm has developed a considerable range of projects, while maintaining a keen awareness of the places where we work and developing thoughtful design solutions. The portfolio of completed works over the last sixteen years, foster our ideals of exceptional design and implementation with tremendous attention to detail.

Our completed works are influenced by the landscapes which surround them. Our goal is to identify unique local characteristics, and incorporate those features with subtle and abstract interpretations. We believe it is possible to successfully integrate and balance practical concerns with art, ecology, environmental sensibilities, and the culture and history of place. Jeffrey Carbo Landscape Architects have won over 40 design awards on the state and national level.

Levy Chamizer Architects

Levy Chamizer Architects is an architecture and interior design studio based in Tel Aviv, Israel. Our diverse portfolio includes trendy retail environments, high-end residential projects, restaurants and showrooms.

The founders graduated from prestigious educational institutions and have experience working abroad. Architect Dolfi Levy graduated from Belgrano University in Argentina. Designer Adi Chamizer graduated from the Fashion Institute of Technology in New York.

Our practice offers a unique fusion of architecture and interior design. Our custom-tailored solutions reach well beyond the conventional architectural package, ranging from the initial conceptual design phase, generating technical drawings, construction supervision and design of particular decorative elements.

We are committed to serve our clients by creating designs and environments that are both functional and aesthetically pleasing. We aim to provide an unparalleled level of service and quality in every project.

L&A Design Group

L&A Design Group is one of the leading and largest companies dedicated to landscape design, urban planning and architectural design in China. L&A was established in 1999 by Mr. Bo Li, a Canadian registered landscape architect, and moved to Shenzhen in the year of 2001. Today, L&A has offices in Shenzhen, Shanghai, Beijing and Xi'an, together they serve clients all over the great regions of China. L & A follows the design principle of landscape urbanism and believes in an integrated approach in the solution of China's complicated urban issues.

Till now L&A has accomplished over 600 reputable landscape, architectural and urban design projects in China with many won national awards. L&A takes pride in its dynamic international design team, with a multi-disciplinary team of more than four hundred professionals including registered architects, landscape architects, planners, engineers, economists and artists.

Maximize Design

Maximize Design is a small garden design office in the heart of Dublin City. Their main focus is to offer practical and sustainable design solutions for city gardens of all sizes.

They have patented their own system of movable planters and green walls, which can be seen in many of their courtyard and rooftop garden designs.

Maximize Design opened its doors in 2008, after the success of the first show garden of head designer Maximilian Kemper at the Bloom in the Park Festival in Dublin, Ireland.

Garden exhibitions and awards include:

Bloom in the Park, Dublin, 2008 – Silver Gilt

Bloom in the Park, Dublin, 2009 - Bronze

Stockseehof Garden Festival, Hamburg, 2010 – Best in Show

Green Machines Exhibition, Science Gallery, Dublin, 2010 – Shortlisted for innovation

Ohtori Consultants Environmental Design Institute

This institute consists of a group of designers specializing in landscape architecture. They deal with all kinds of environmental design including landscape planning and architectural design, and always strive to propose and realize the creation of new cities and regional environment incorporating a variety of fields from engineering, landscape gardening, and ecology into our design fields. The content of their business ranges from urban development, commercial space, plaza, living environment, green land of park to garden design. In recent years, they have been engaged in many overseas projects as well.

ONG&ONG Pte Ltd.

With a track record of 40 years in the industry, ONG&ONG has earned an unparalleled reputation for servicing our clients with creativity, excellence and commitment.

ONG&ONG offers a complete 360° solution – i.e. a parceled cross-discipline integrated solution, encompassing all aspects of the construction business. This three-pronged solution encompasses design (architecture, urban planning, interior, landscape, environmental branding, lighting and experience design), engineering (mechanical, electrical, civil, structural, fire safety and environmental) and management (project, development, construction, cost and place).

We are an ISO14001 certified practice with offices in Singapore, China, Vietnam, India, Malaysia, the USA, Indonesia as well as Mongolia. In-depth knowledge of the local context, culture and regulations allow us to better understand our clients' needs to enable us to meet and exceed their expectations. With several awards to our name, ONG&ONG constantly works towards staying ahead of the competition.

Peter Glass & Associates

Peter Glass & Associates is one of the Australia's longest established and most experience landscape architecture practices.

The team of eight specialist landscape architects and designers, plus additional support staff, is proud of their fine reputation, earned by consistently achieving excellence in both design and documentation, and by providing the highest level of service. This has enabled us to assist numerous public, private and corporate clients, both in Australia and overseas, for over three decades. Their areas of expertise include Landscape Architecture (residential, commercial and public), Swimming Pool and Water Feature Design, Environmental Planning and Urban Design, Expert Witness and Testimony.

Their extreme pride in their work, combined with innovative design, thorough documentation, attention to detail and an awareness of commercial realities, ensures that their experienced team of landscape architects and designers at Peter Glass and Associates produce projects of only the highest quality. They have built reputation on providing their clients with experienced, personalised and cost effective services encompassing private residential pool and landscape design, commercial and public domain design and expert witness and testimony.

PLACE Design Group Pty Ltd.

Place Design Group (PDG) is a leading planning, design, and environment consultancy, with offices in major international markets including Australia, Asia, the Middle East, and the Pacific. In an environment where rhetoric often rules, the collaborative that is PLACE delivers intelligent, creative, and sustainable solutions based on a thorough understanding of client objectives and the contemporary built environment.

Place recognizes the advantages that a comprehensive service can offer by delivering an exceptional level of project integration by managing more of the development continuum in-house. Our high performing teams are dedicated to delivering fresh and innovative, best-practice development outcomes, underpinned by a cultural commitment to creating great places.

Rios Clementi Hale Studios

RIOS CLEMENTI HALE STUDIOS

Rios Clementi Hale Studios, established in 1985, has developed an international reputation for its collaborative and multi-disciplinary approach, establishing an award-winning tradition across an unprecedented range of design disciplines. Acknowledging the firm's varied body of work, the American Institute of Architects California Council gave Rios Clementi Hale Studios its 2007 Firm Award, the organization's highest honor. For its varied landscape work—from civic parks to private gardens—the firm was named a finalist in the 2009 National Design Awards. The architecture, landscape architecture, planning, urban, interior, exhibit, graphic, and product designers at Rios Clementi Hale Studios create buildings, places, and products that are thoughtful, effective, and beautiful.

Rosmarin landscape Design

Rosmarin Landscape Design is a company dedicated to creating gardens of individuality and atmosphere, has expanded the business to Sonoma County, California and continues to develop projects in the USA, South Africa and the UK. Their gardens are artistically designed and professionally installed. Local horticultural experience and an international art and garden background combine to create an aesthetically satisfying, functional and cost effective result.

Their strengths are visualizing the garden at various times throughout the year and from different points of view as a series of painterly canvases while never losing sight of the structure of the overall design and the practical uses defined by the people who live in it. Drawing from their experiences in Europe and Africa, our trademarks combine informal, full, naturalistic planting that mirrors the contour of the natural landscape with classic architectural lines for more formal designs.

A garden for all seasons is their goal.

Sasaki Associates, Inc.

Sasaki was founded 60 years ago on the basis of interdisciplinary planning and design. Today, our services include architecture, interior design, planning, urban design, landscape architecture, strategic planning, civil engineering, and graphic design. Among these disciplines, we collaborate with purpose. Our integrated approach yields rich ideas and surprising insights.

We approach our work from a foundation of wide-ranging expertise and bring fresh energy and innovation to each project. Our culture is inquisitive—we are passionate about ideas. Our professionals embark on research efforts and contribute to thought leadership in our respective disciplines.

From our studio in Boston, Massachusetts, we work in a variety of settings—locally, nationally, and globally. We ask the right questions and listen attentively. We are team-based, both internally and with our clients. Together, we examine the problem and the context in which it exists.

Sasaki is an innovator. We play a leading role in shaping the future of the built environment through bold ideas and new technologies. We approach sustainability through the lenses of economics, social context, and the environment. Our solutions are not only effective—they are poetic and enduring. Our approach helps clients make smart, long-term decisions that result in greater value for them, and a better future for the planet.

Silvia Glaßer und Udo Dagenbach

Silvia Glasser and Udo Dagenbach founded their partnership in 1988. Since then they created innovative and highest quality parks and landscapes, but also especially public places and private gardens. Silvia Glasser is a state approved gardener specializing in perennials and received her Diploma in Landscape Architecture by the University of Nuertingen in 1985. Udo Dagenbach is a state approved gardener too with a Diploma in Landscape Architecture from the Technical University of Berlin in 1986. He also studied stone sculpture at the University of Art in Berlin as a guest student of professor Makoto Fujiwara and worked with him. The small office with three landscape architects and two to three assistants primarily focuses on the new design of public parks, Hotels and Resorts and private residential projects as well as the reconstruction of gardens and parks. Their philosophy: The settings of gardens and parks form backdrops before which visitors, whether public or private, are able to act out a role in their very own play.

The projects are spread over Europe and Eastern Europe.

SWA Group

For over five decades, SWA Group has been recognized as a world design leader in landscape architecture, planning and urban design. Our projects have received over 600 awards and have been showcased in over 60 countries. Our principals are among the industry's most talented and experienced designers and planners. Emerging in 1959 as the West Coast office of Sasaki, Walker and Associates, the firm first assumed the SWA Group name in 1975.

Despite being one of the largest firms of its type in the world, SWA is organized into smaller studio-based offices that enhance creativity and client responsiveness. Over 75% of our work has historically come from repeat clients. In addition to bringing strong aesthetic, functional, and social design ideas to our projects, we're also committed to integrating principles of environmental sustainability. At the core of our work is a passion for imaginative, solution-oriented design that adds value to land, buildings, cities, regions, and to people's lives.

Terragram Pty Ltd

Whilst the initial impetus for Terragram's existence came from winning a competition (1985), another incentive was to have a working platform that would engage in a critique of the then prevailing pragmatic approach to designed landscape, pursued by most practices in Australia in the mid eighties. Terragram is not a traditional office. The French word 'atelier' better describes the atmosphere and inclination to experiment, cherish intuitive responses, invite the unexpected and scary, and to technologically innovate. Despite the modest size of the company, interest range also includes stage design, sculpture, graphics, furniture and educational activities. Cultural curiosity has led Terragram to work in different countries outside Australia. (Bolivia, China, Czech Republic, France, Germany, Indonesia, Iran, Israel, Italy, Japan, Korea, New Caledonia, Singapore, Switzerland, Taiwan, the United States of America and Vietnam).